Recent Titles in This Series

503 **Michael Makkai,** Duality and definability in first order logic, 1993
502 **Eriko Hironaka,** Abelian coverings of the complex projective plane branched along configurations of real lines, 1993
501 **E. N. Dancer,** Weakly nonlinear Dirichlet problems on long or thin domains, 1993
500 **David Soudry,** Rankin-Selberg convolutions for $SO_{2\ell+1} \times GL_n$: Local theory, 1993
499 **Karl-Hermann Neeb,** Invariant subsemigroups of Lie groups, 1993
498 **J. Nikiel, H. M. Tuncali, and E. D. Tymchatyn,** Continuous images of arcs and inverse limit methods, 1993
497 **John Roe,** Coarse cohomology and index theory on complete Riemannian manifolds, 1993
496 **Stanley O. Kochman,** Symplectic cobordism and the computation of stable stems, 1993
495 **Min Ji and Guang Yin Wang,** Minimal surfaces in Riemannian manifolds, 1993
494 **Igor B. Frenkel, Yi-Zhi Huang, and James Lepowsky,** On axiomatic approaches to vertex operator algebras and modules, 1993
493 **Nigel J. Kalton,** Lattice structures on Banach spaces, 1993
492 **Theodore G. Faticoni,** Categories of modules over endomorphism rings, 1993
491 **Tom Farrell and Lowell Jones,** Markov cell structures near a hyperbolic set, 1993
490 **Melvin Hochster and Craig Huneke,** Phantom homology, 1993
489 **Jean-Pierre Gabardo,** Extension of positive-definite distributions and maximum entropy, 1993
488 **Chris Jantzen,** Degenerate principal series for symplectic groups, 1993
487 **Sagun Chanillo and Benjamin Muckenhoupt,** Weak type estimates for Cesaro sums of Jacobi polynomial series, 1993
486 **Brian D. Boe and David H. Collingwood,** Enright-Shelton theory and Vogan's problem for generalized principal series, 1993
485 **Paul Feit,** Axiomization of passage from "local" structure to "global" object, 1993
484 **Takehiko Yamanouchi,** Duality for actions and coactions of measured groupoids on von Neumann algebras, 1993
483 **Patrick Fitzpatrick and Jacobo Pejsachowicz,** Orientation and the Leray-Schauder theory for fully nonlinear elliptic boundary value problems, 1993
482 **Robert Gordon,** G-categories, 1993
481 **Jorge Ize, Ivar Massabo, and Alfonso Vignoli,** Degree theory for equivariant maps, the general S^1-action, 1992
480 **L. Š. Grinblat,** On sets not belonging to algebras of subsets, 1992
479 **Percy Deift, Luen-Chau Li, and Carlos Tomei,** Loop groups, discrete versions of some classical integrable systems, and rank 2 extensions, 1992
478 **Henry C. Wente,** Constant mean curvature immersions of Enneper type, 1992
477 **George E. Andrews, Bruce C. Berndt, Lisa Jacobsen, and Robert L. Lamphere,** The continued fractions found in the unorganized portions of Ramanujan's notebooks, 1992
476 **Thomas C. Hales,** The subregular germ of orbital integrals, 1992
475 **Kazuaki Taira,** On the existence of Feller semigroups with boundary conditions, 1992
474 **Francisco González-Acuña and Wilbur C. Whitten,** Imbeddings of three-manifold groups, 1992
473 **Ian Anderson and Gerard Thompson,** The inverse problem of the calculus of variations for ordinary differential equations, 1992
472 **Stephen W. Semmes,** A generalization of riemann mappings and geometric structures on a space of domains in $\mathbf{C}^n$, 1992

(*Continued in the back of this publication*)

MEMOIRS
of the
American Mathematical Society

Number 503

Duality and Definability in First Order Logic

Michael Makkai

September 1993 • Volume 105 • Number 503 (fourth of 6 numbers) • ISSN 0065-9266

American Mathematical Society
Providence, Rhode Island

1991 *Mathematics Subject Classification*.
Primary 03C20, 03C40, 03G30, 18D05.

Library of Congress Cataloging-in-Publication Data

Makkai, Michael, 1933–
Duality and definability in first order logic/Michael Makkai.
p. cm. – (Memoirs of the American Mathematical Society; no. 503)
Includes bibliographical references.
ISBN 0-8218-2565-8
1. First-order logic. 2. Duality theory (Mathematics) 3. Toposes. I. Title. II. Series.
QA3.A57 no. 503
[QA9]
510 s–dc20 93-4868
[511.3] CIP

Memoirs of the American Mathematical Society

This journal is devoted entirely to research in pure and applied mathematics.

Subscription information. The 1993 subscription begins with Number 482 and consists of six mailings, each containing one or more numbers. Subscription prices for 1993 are $336 list, $269 institutional member. A late charge of 10% of the subscription price will be imposed on orders received from nonmembers after January 1 of the subscription year. Subscribers outside the United States and India must pay a postage surcharge of $25; subscribers in India must pay a postage surcharge of $43. Expedited delivery to destinations in North America $30; elsewhere $92. Each number may be ordered separately; *please specify number* when ordering an individual number. For prices and titles of recently released numbers, see the New Publications sections of the *Notices of the American Mathematical Society.*

Back number information. For back issues see the *AMS Catalog of Publications.*

Subscriptions and orders should be addressed to the American Mathematical Society, P. O. Box 1571, Annex Station, Providence, RI 02901-1571. *All orders must be accompanied by payment.* Other correspondence should be addressed to Box 6248, Providence, RI 02940-6248.

Memoirs of the American Mathematical Society is published bimonthly (each volume consisting usually of more than one number) by the American Mathematical Society at 201 Charles Street, Providence, RI 02904-2213. Second-class postage paid at Providence, Rhode Island. Postmaster: Send address changes to Memoirs, American Mathematical Society, P. O. Box 6248, Providence, RI 02940-6248.

Printed in the United States of America.
This volume was printed directly from author-prepared copy.
The paper used in this book is acid-free and falls within the guidelines
established to ensure permanence and durability. ♾
Printed on recycled paper. ♻
10 9 8 7 6 5 4 3 2 1 98 97 96 95 94 93

TABLE OF CONTENTS

ABSTRACT

We develop a duality theory for small Boolean pretoposes in which the dual of T is the groupoid of models of a Boolean pretopos T equipped with additional structure derived from ultraproducts. The duality theorem says that any small Boolean pretopos is canonically equivalent to its double dual. We use a strong version of the duality theorem to prove the so-called descent theorem for Boolean pretoposes which says that category of descent data derived from a conservative pretopos morphism between Boolean pretoposes is canonically equivalent to the domain-pretopos. The descent theorem contains the Beth definability theorem for classical first order logic. Moreover, it gives, via the standard translation from the language of categories to symbolic logic, a new definability theorem for classical first order logic concerning set-valued functors on models, expressible in purely syntactical (arithmetical) terms .

Key words and phrases: pretopos, first order logic, duality theory, definability theory, ultraproduct, category of models, descent theory, 2-category, exactness property.

INTRODUCTION

The main aim of this paper is to prove the so-called descent theorem for Boolean pretoposes. In section 2, the theorem will be stated in categorical terms, and in section 3, in the language of symbolic logic. In its symbolic-logical form, the theorem is an apparently new result for pure first order logic. In fact, it is a statement on the syntax of first order logic, whose arithmetic complexity is Π^0_2, similarly to the Beth definability theorem (see [C/K]), and to many other model-theoretical results (syntactic characterizations, preservation theorems) that may be stated in a purely syntactical manner. (The Π^0_2-form comes from the fact that the results in question, including the theorem of this paper, are of the general form: "for all deductions of a certain kind, there is another deduction of a certain other kind", where the "kinds" in question are given by recursive conditions).

The descent theorem contains the Beth definability theorem as a part, and it may be considered as a definability theorem for implicitly definable new primitives that are not necessarily subsets of the basic universe, but may be added on the outside, on new sorts. (H. Gaifman's theorem (see e.g.[M3]), which is a definability theorem of a similar general character, is not the same as the descent theorem. As we will see, Gaifman's result can be deduced from the present work; I do not see how to obtain the main theorem of this paper from Gaifman's theorem, or his methods.) The descent theorem is a statement concerning the definability properties of set-valued functors on certain categories of models.

Let me give the main points of the history of the result. The theorem is, in the first place, inspired by, and formally analogous to, the descent theorem for open geometric morphisms established by A. Joyal and M. Tierney in [J/T] , a paper of fundamental importance for, among others, categorical logic. That paper contains the discovery of a far-reaching analogy between the "algebra" of (infinitary) first order logic, and "Abelian" algebra. In fact, the descent theorems of [J/T] are analogous to A. Grothendieck's descent theorem [G] for modules, and sheaves of modules.

It was A. M. Pitts who conjectured the theorem of this paper in the first place, alongside others to be mentioned below, in the context of his work involving a transfer of results and "spirit" from [J/T] to finitary logic (see [P1], [P2], [P3], [P4]). The mechanism of the transfer, taking the shape of functorial constructions, is Pitts' main discovery; it leads him to new results on finitary logic, as well as to the most satisfactory treatment available of interpolation and definability in the usual sense for intuitionistic logic.

In his thesis [Z1], M. Zawadowski proved one of Pitts' conjectures, the lax descent theorem for pretoposes; see also [Z2]. Against my initial skepticism, he started out on his way to the proof with a plan of applying my duality theory [M1]. He successfully completed his plan, and contributed, among others, a highly surprising and beautiful argument, which, suitably transformed, plays a crucial role in this paper as well.

This paper is the result of trying to repeat Zawadowski's feat for the Boolean descent theorem. The proof, in fact, follows Zawadowski's outline quite faithfully. On the other hand, there are two essentially new features.

One is that my original duality theory, serving Zawadowski's purposes, had to be replaced by another one. The new duality theory for Boolean pretoposes is given in sections 4 - 8. It builds on the old theory for pretoposes in general, but it also involves further complexities, most visible in Section 6 on the "syntax of special ultramorphisms". It turns out that the technical notions of "cell-system" and the like, brought out explicitly in [M2], but appearing implicitly in [M1] already, are useful in the context of this paper too. They are subjected to a manipulation which is the main technical contribution in this paper.

The other feature is the "preparation of functor specifications" (Section 9) for the treatment which is the analog of Zawadowski's main argument. This preparation reduces, in Zawadowski's proof, to an essentially trivial, although important fact (pointed out by Pitts). The main point in Section 9 is Lemma 1, a forcing argument, "forcing with generic automorphisms", which, in a somewhat different form, formed a part of an unpublished piece of work done by M. Ajtai and myself in 1979. The argument seems to be quite fundamental, and I would not be surprised if in the meantime it had appeared in the literature in some form.

It is well-known that the Beth definability theorem can be proved in an elementary way; more precisely, within (first order, even recursive) arithmetic. The question remains whether the descent theorems (the one for pretoposes, and the one for Boolean pretoposes) can be so proved. Let us mention that Pitts' third conjecture, the descent theorem for Heyting pretoposes, is still open.

In the first section, we will go through a thoroughly model-theoretical (in the sense of the model theory of propositional logic, in the style of Section 1.2 of [C/K]), and at the same time categorical, proof of the Beth definability theorem for classical propositional logic. The very statement of our main theorem, and later its proof as well, will be obtained by guessing proper generalizations of the propositional situation. The result will not be the classical Beth definability theorem for predicate logic, but something considerably (it seems) stronger.

The basis for the possibility of such a generalization is the fundamental, and not sufficiently appreciated, fact that the notion of category is a generalization of the notion of partial order, or even preorder: in fact, a preorder is nothing but a category in which every hom-set is of cardinality at most 1 (" 2-enriched category"). The fundamental notions for the theory of lattices and Boolean algebras, infs (greatest lower bounds) and sups (least upper bounds) of families of elements are generalized in the notions of (projective) limits, and colimits (inductive limits), the bread and butter of category theory. A basic strategy underlying this paper (in this respect, the paper is certainly not alone!) is the lifting of facts and constructions from posets to categories.

A further parallel to be exploited, possibly despite our initial disbelief, is one between the 2-element total ordering 2 , and `Set` , the category of (small) sets and functions. The fact that there is a fruitful parallel between those two objects is well-known in category theory; e.g., the theory of profunctors is based on such a consideration. However, the way we exploit the parallel is not along the lines of general category theory: what is happening in this paper is *pure* category theory, but not *general* category theory.

Classical propositional logic may be identified with the study of the properties of 2 endowed with the operations of finite infs and sups, as well as complementation, resulting in the theory of Boolean algebras. Categorical logic shows (see, e.g., [M/R], also [M3]) that classical first order predicate logic can be regarded as the study of finite limits and (certain) finite colimits (along with complementation of subobjects) within `Set` , resulting in the notion of (Boolean) pretopos, due to A. Grothendieck [SGA4]. For instance, as the completeness theorem for propositional logic (Emil Post, 1921) is expressed in the theory of Boolean algebras as the Stone representation theorem, the Gödel(/Malcev) completeness theorem for first order logic is "translation"-equivalent to the Deligne(/Joyal) representation theorem [SGA4] for pretoposes, a result of great formal similarity to the Stone result. It is precisely the *formal* similarities that interest us in the first place. We will try to guess new results in predicate logic by formally lifting known situations from the propositional case, via the categorical framework, and in fact, we will even try to guess the new proofs in this way. It so happens that we can do much along these lines, and the enterprise takes us on a journey in interesting new mathematics.

Another aspect of our lifting strategy is the passing from *categories* of the basic structures involved (category of Boolean algebras, category of Stone spaces) to *2-categories* of the new basic structures (pretoposes, ultracategories). Every time one deals with category-based structures, those structures will form, most naturally, 2-categories (or possibly bicategories, which are more complex than 2-categories). The universal algebra of category-based structures largely overlaps with 2-(bi-)category theory.

Nevertheless, in the concluding section, which contains the finishing touches to the proof of the descent theorem, I will give a version of the proof that is stripped of the 2-category theory. Sections 2 and 11 will thus be rendered superfluous, at the expense, however, of losing the conceptual formulation of the theorem as Theorem 2.2 in Section 2, as well as losing much of the heuristics underlying the work.

Of course, there is an obligatory disclaimer called for here: I am not sure that the descent theorem really requires the machinery used and developed here. Even if the answer is "no", however, I am pleased with the motivation it has given me to look for the duality theorem for Boolean pretoposes, which I had long thought ought to look something like the theorem here

(Theorem 8.1), but which I would hardly have taken the trouble to figure out without the prodding of an application.

I have made an effort to make the paper reasonably self-contained. The necessary background for reading the paper consists of basic notions of category theory (as can be found in [CWM]), and the connections established between first order logic and pretoposes in [M/R]. The references [B/J] and [M3] may also be consulted for the latter. 2-categories are mentioned in some abstract formulations, but in fact, only some very "concrete" 2-categories are actually used. The reference [M/Pa] contains a certain amount of 2-category theory, tailored for needs related to logic, certainly sufficient for the present purposes. At some points, technical matters from [M1] and [M2] are used.

The set-theoretic foundations adopted are those using Grothendieck universes (see [SGA4]), also explained in section 1.1 of [M/Pa], except that what were written $\mathbf{U}_1$, $\mathbf{U}_2$, $\mathbf{U}_3$ there are called $\mathcal{U}_0$, $\mathcal{U}_1$, $\mathcal{U}_2$ now.

The numbering of items such as definitions and lemmas starts from the beginning in each section; when an item is referred to in another section, the number of the item's section is prefixed to the item number.

I express my thanks to Marek Zawadowski for many fruitful conversations, and for his decisive insights in the subject at hand. I also thank George Janelidze, Bill Boshuck and the referee for their valuable remarks, which have contributed to improvements of the presentation.

1. BETH'S THEOREM FOR PROPOSITIONAL LOGIC

In any category that is sufficiently complete and cocomplete (precisely, that has pullbacks and coequalizers of equivalence relations), we may consider the *regular factorization* of an arrow $f: A \longrightarrow B$ as follows: take first

$$K \underset{p_1}{\overset{p_0}{\rightrightarrows}} A \xrightarrow{\;f\;} B \ ,$$

the kernel-pair of f (that is, $\begin{array}{ccc} A & \xrightarrow{f} & B \\ {\scriptstyle p_1}\uparrow & & \uparrow{\scriptstyle f} \\ K & \xrightarrow[p_2]{} & A \end{array}$ is a pullback-square); then take

$$K \underset{p_1}{\overset{p_0}{\rightrightarrows}} A \xrightarrow{\;s\;} C \, ,$$

the quotient (coequalizer) of the pair (p_0, p_1) ; in the commutative diagram

$$K \underset{p_1}{\overset{p_0}{\rightrightarrows}} A \xrightarrow{\qquad f \qquad} B \ , \qquad A \xrightarrow{\;s\;} C \xrightarrow{\;i\;} B$$

i is given by the universal property of s ; the factorization $f = i \circ s$ so obtained is the *regular factorization* of f . It is determined up to isomorphism (in a straightforward precise sense), since the ingredients used, all defined by universal properties, are so determined.

In $\mathtt{Set}$, the category of (small) sets and functions, the regular factorization is the same as the surjective/injective factorization. Many concrete categories (categories equipped with a conservative functor to $\mathtt{Set}$) *inherit* this property from $\mathtt{Set}$; e.g., the ones that are monadic over $\mathtt{Set}$, regular-epi-reflective subcategories of the latter, etc. (the reason is the fact that epis split in $\mathtt{Set}$).

We will talk about the *coregular factorization* of an arrow in a category when we mean the regular factorization of the corresponding arrow in the opposite category. We will show that

---

Received by the editor February 11, 1991 and, in revised form, April 7, 1992.

Supported by NSERC Canada and FCAR Quebec.

the coregular factorization in Boole *, the category of Boolean algebras is the standard surjective/injective factorization.*

Before doing so, however, let's see the connection with the Beth theorem. Let L be a language (in the sense of [C/K]) for propositional logic, that is, a set of atomic propositions. Let P be a further atom, and $L(P) = L \cup \{P\}$. Let T be a theory in the language $L(P)$, that is, a set of $L(P)$-sentences. We write $T(P)$ for T; replacing P by a new atom Q throughout results in $T(Q)$. We say that T *implicitly defines* P over L, if $T(P) \cup T(Q) \vdash P \longleftrightarrow Q$. Here, $\vdash$ denotes formal deducibility; via completeness, implicit definability means that any L-structure (interpretation of all atoms in L by **true** or **false**) can be completed in at most one way to a model of T. T *explicitly defines* P over L if there is an L-sentence φ such that $T(P) \vdash P \longleftrightarrow \varphi$. The theorem says that

if T *implicitly defines* P *over* L *, then* T *explicitly defines* P *over* L .

The converse is obvious.

With any theory T in any language L (in propositional logic), we associate $\mathrm{LT}(L, T)$, the *Lindenbaum-Tarski algebra of* (L, T). $\mathrm{LT}(L, T)$ is the Boolean algebra whose elements are the equivalence classes $\varphi/\sim$ of L-sentences φ under the equivalence $\varphi \sim \psi \iff T \vdash \varphi \longleftrightarrow \psi$; the partial order of $\mathrm{LT}(L, T)$ is $\varphi/\sim \,\le\, \psi/\sim \iff T \vdash \varphi \to \psi$. Let, with our notation in the Beth theorem, $A = \mathrm{LT}(L, \emptyset)$ (empty theory over L; free Boolean algebra over the generators in L), $B = \mathrm{LT}(L(P), T)$, and $D = \mathrm{LT}(L \cup \{P, Q\}), T(P) \cup T(Q))$. We have obvious Boolean homomorphisms $f: A \longrightarrow B$, $B \overset{g}{\underset{h}{\rightrightarrows}} D$: $f(\varphi/\sim_0) = \varphi/\sim_1$, $g(\varphi/\sim_1) = \varphi/\sim_2$, $h(\varphi/\sim_1) = \varphi[Q/P]/\sim_2$, where $\sim_0$, $\sim_1$, $\sim_2$ are the equivalences used to define A, B and D, respectively, and $\varphi[Q/P]$ is obtained by substituting Q for P. We claim that the square

$$\begin{array}{ccc} B & \xrightarrow{\;g\;} & D \\ {\scriptstyle f}\uparrow & & \uparrow{\scriptstyle h} \\ A & \xrightarrow[\;f\;]{} & B \end{array}$$

is a pushout. The verification of this claim uses only the general features of the concepts involved, and it is an easy calculation.

Let's take the equalizer $i: C \to B$ of (g, h), and the coregular factorization of f:

$$\begin{array}{ccccc} A & & \xrightarrow{\quad f \quad} & & B \overset{g}{\underset{h}{\rightrightarrows}} D. \\ & {\scriptstyle s}\searrow & & \nearrow{\scriptstyle i} & \\ & & C & & \end{array}$$

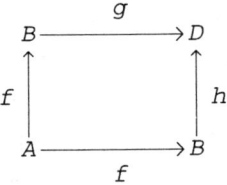

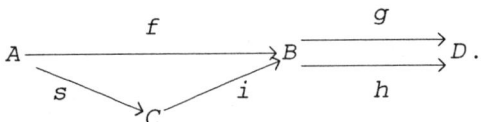

By the factorization theorem, s is surjective. Now, assume that T implicitly defines P over L. Then, by definition, for $x = P/{\sim_1}$, $g(x) = P/{\sim_2} = Q/{\sim_2} = h(x)$. Since i is an equalizer, there is a (unique) $y \in E$ such that $i(y) = x$, and since s is surjective, there is $z = \varphi/{\sim_0} \in A$, φ an L-formula, such that $s(z) = y$, hence $f(z) = is(z) = x$, which is to say that $T \vdash \varphi \longleftrightarrow P$ as promised.

An appropriate "PC-"generalization of Beth (of a kind that is well-known in model theory) can in fact be shown to be *equivalent*, via the process of translation from propositional theories and Boolean algebras and back, to the factorization theorem for Boolean algebras.

Our proof of the factorization theorem will be an application of the Stone duality theory. The idea of the proof is simple; the duality theory presents the opposite of Boole as a certain well-understood category Stone, that of Stone spaces. Since the latter is sufficiently algebraic, the regular factorization in it is the standard one. By duality, the regular factorization in Stone corresponds to the coregular factorization in Boole. It takes a little further argument to conclude the result.

We will state the Stone theory to the extent, and in the form, that we need it. In particular, no topology will be involved; this is important for the 2-categorical generalization we use for the main purpose.

A *concrete category* is just a category $\mathbf{C}$ with a functor (called "forgetful") to Set. Many categories in practice appear as concrete categories, because a natural forgetful functor is present. One usually makes assumptions about the forgetful functor (e.g. that it is faithful); we will state those explicitly as we need them. An arrow in a concrete category is *surjective* or *injective* if its image under the forgetful functor in Set is a surjective or injective function, respectively.

The *Zawadowski setup*, by definition, consists of an adjunction between two concrete categories

$$\mathbf{A}^{op} \underset{G}{\overset{F}{\rightleftarrows}} \mathbf{S}, \quad F \dashv G \quad (\varepsilon : \text{counit})$$

with the following conditions imposed:

Z0. $\mathbf{A}$ and $\mathbf{S}$ are finitely complete and cocomplete, the forgetful functor on $\mathbf{A}$ preserves equalizers and reflects isomorphisms.

Z1. $\varepsilon_A : A \longrightarrow FGA$ (in $\mathbf{A}$) is injective for all $A \in \mathbf{A}$.

Z2. $$\frac{X \longrightarrow GA}{A \longrightarrow FX}\text{(transpose)} \qquad \begin{array}{c}\text{injective}\\ \downarrow\\ \text{surjective}\end{array};$$

that is, the transpose of an injective arrow in $\mathbf{S}$ is surjective in $\mathbf{A}$.

Z3. The regular factorization of an arrow of the form Gf in $\boldsymbol{S}$, with any arrow f in $\boldsymbol{A}$, is the surjective/injective factorization.

1. Proposition. Suppose a Zawadowski setup $(\boldsymbol{A},\ldots)$. Then the coregular factorization in $\boldsymbol{A}$ is the surjective/injective factorization.

Proof. Note, first of all, that Z2 applied to $GA \xrightarrow{1} GA$, gives that ε_A is surjective; hence, by Z1, it is an isomorphism (the forgetful functor on $\boldsymbol{A}$ reflects iso's). This means that the Zawadowski setup presents the category $\boldsymbol{A}^{\mathrm{op}}$ as a full reflective subcategory of $\boldsymbol{S}$, among others. Let's write * for the actions of both F and G. Start with

$$A \xrightarrow{f} B \quad \text{in } \boldsymbol{A};$$

construct

$$A \xrightarrow{f} B \rightrightarrows K, \tag{1}$$

the cokernel pair of f. Apply G:

$$A^* \longleftarrow B^* \leftleftarrows K^*;$$

we get the kernel pair of f^*, since G as a right adjoints preserves limits (takes colimits in $\boldsymbol{A}$ to limits in $\boldsymbol{S}$). Take the regular factorization in $\boldsymbol{S}$:

$$\begin{array}{ccccc} A^* & \longleftarrow & B^* & \leftleftarrows & K^* \\ & {\scriptstyle i}\nwarrow & \swarrow & & \\ & & S & & \end{array} \quad . \tag{2}$$

Apply F:

$$\begin{array}{ccccc} A^{**} & \xrightarrow{f^{**}} & B^{**} & \rightrightarrows & C^{**} \\ & \searrow & \nearrow{\scriptstyle e} & & \\ & & S^* & & \end{array} \quad . \tag{3}$$

By reflection, this is essentially the same as

$$\begin{array}{ccccc} A & \xrightarrow{f} & B & \rightrightarrows & C; \\ & \searrow & \nearrow & & \\ & & S^* & & \end{array}$$

more precisely, we have

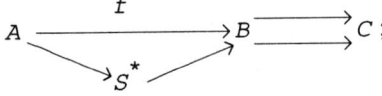

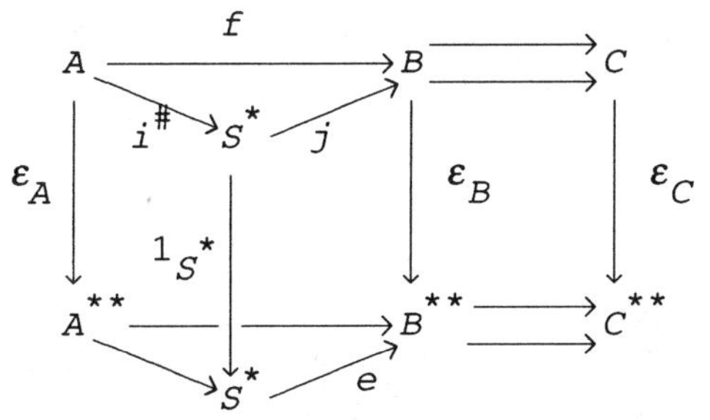

where the upper horizontal part of the diagram is the same as (1), the lower part the same as (3), the vertical arrows are isomorphisms, and $i^{\#}$ and j are defined so that the corresponding quadrangles commute, making $i^{\#}$ the transpose of i. By the naturality of ε, two quadrangles on the right commute, and also, $f = j \circ i^{\#}$. Since F as a left adjoint preserves colimits, e is the equalizer of the two arrows out of its codomain. Hence, j is an equalizer of the two arrows out of its codomain as well, and the upper half is the coregular factorization of f.

j is automatically injective (since the forgetful functor $\mathbf{A} \longrightarrow \mathrm{Set}$ preserves equalizers). i is injective by Z3. Hence, by Z2., $i^{\#}$ is surjective. [] 1.

As we said in the above proof, $\mathbf{A}^{\mathrm{op}}$ is essentially a full subcategory of $\mathbf{S}$. Since we are proving something about the category $\mathbf{A}$, we should ask to what extent the *larger* category $\mathbf{S}$ is used in the proof. The answer is that the object S that appears in (2) is the one (and only one) object *possibly outside* $\mathbf{A}$ that is used in the proof. Below, we apply the setup to Boolean algebras as $\mathbf{A}$; this application will however be redundant in the sense that in that case, the object $\mathbf{S}$ will, in fact, be in (the image of) $\mathbf{A}$. On the other hand, in our main application of a 2-categorical version of the setup, (the analog of) the object S will definitely not (aside from extreme cases for the initial data $f: A \longrightarrow B$) come from $\mathbf{A}$.

2. Proposition. The category Boole of Boolean algebras is part of a Zawadowski setup (Boole,...).

Proof (sketch). An *ultrafilter* (I, U) is a Boolean homomorphism $2^I \longrightarrow 2$, from the ordinary I-th power 2^I of the 2-element Boolean algebra 2, to 2 itself. We let L be the (large) similarity type consisting of all ultrafilters (I, U), each (I, U) meant to be an I-ary operation symbol. An algebra of type L is an entity $S = (|S|, \langle U^{(S)} \rangle_{(I, U) \text{ any u.f.}})$, with $|S|$ a small set, and with $U^{(S)}: |S|^I \longrightarrow |S|$, an I-ary operation on $|S|$ for each ultrafilter (I, U). $\mathbf{S}$ is defined to be the category whose objects are the algebras of type L, and whose arrows are the L-homomorphisms.

The forgetful functors on $\mathtt{Boole}$ as well as on $\boldsymbol{S}$ are taken to be the usual underlying-set functors. Condition Z0 is obviously satisfied.

There is a specific object in $\boldsymbol{S}$, $\mathbf{2}$; its underlying set $|\mathbf{2}| = 2$, and its operations are $U^{(\mathbf{2})} = U$, one for each ultrafilter (I, U). If $S \in \boldsymbol{S}$, then $\hom(S, \mathbf{2}) = \hom_{\boldsymbol{S}}(S, \mathbf{2})$ is a subset of $|2^{|S|}|$, the underlying set of the power $2^{|S|}$ of the 2-element Boolean algebra with exponent $|S|$. In fact, $\hom(S, 2)$ is closed under the operations of $2^{|S|}$, as is immediately realized upon considering the connection between the structures 2 and $\mathbf{2}$. Thus, we may consider the subalgebra of the Boolean algebra $2^{|S|}$ with underlying set $\hom(S, \mathbf{2})$; let us call it $\mathit{hom}(S, \mathbf{2})$. For $f: S \longrightarrow S'$ in $\boldsymbol{S}$, the Boolean homomorphism $f^* = 2^f : 2^{|S'|} \longrightarrow 2^{|S|}$ restricts to a Boolean homomorphism

$$\mathit{hom}(f, \mathbf{2}) : \mathit{hom}(S', \mathbf{2}) \to \mathit{hom}(S, \mathbf{2}) .$$

We have defined a contravariant functor

$$F = \mathit{hom}(S, \mathbf{2}) \ : \ \boldsymbol{S} \longrightarrow \mathtt{Boole}^{\mathrm{op}} .$$

We define

$$G = \mathbf{hom}(-, 2) \ : \ \mathtt{Boole}^{\mathrm{op}} \longrightarrow \boldsymbol{S}$$

similarly. For A a Boolean algebra, $\mathbf{hom}(A, 2)$ is the subalgebra of the power-algebra $\mathbf{2}^{|A|}$ in $\boldsymbol{S}$ with underlying set $\hom(A, 2)$.

It is easy to see that F is a left adjoint to G; the counit and unit maps are evaluations:

$$\begin{array}{llll} \varepsilon_A : A & \longrightarrow \mathit{hom}(\mathbf{hom}(A, 2), \mathbf{2}) & & \\ \quad a & \longmapsto [u \longmapsto u(a)] & (u \in \mathbf{hom}(A, 2)) , \end{array}$$

$$\begin{array}{llll} \eta_S : S & \longrightarrow \mathbf{hom}(\mathit{hom}(S, \mathbf{2}), 2) & & \\ \quad s & \longmapsto [x \longmapsto x(s)] & (x \in \mathit{hom}(S, 2)) . \end{array}$$

In fact, $\mathbf{hom}(A, 2)$, with any Boolean algebra A, is a disguised form of the Stone space of A; the infinitary operations $U^{(S)}$, for $S = \mathbf{hom}(A, 2)$, are ultrafilter-limits familiar in the context of compact Hausdorff spaces. Condition Z1 is just the Stone representation theorem, and Z2 can be obtained by an argument similar to one used to prove that the topological version of ε_A is surjective. The proofs are no harder than the ones in the familiar topological formulation.The proof of Z2 is also very similar to the proof of (*) in Section 8 below.

The point of the present formulation, in particular, the choice of the category $\boldsymbol{S}$, is that Z3 becomes obvious; in fact, the regular factorization of all arrows in $\boldsymbol{S}$ is the surjective/injective factorization. This is clear on the basis of the "algebraic" character of $\boldsymbol{S}$.

[To be sure, $\mathcal{S}$ is not monadic over $\mathtt{Set}$ (because of the largeness of the similarity type L), but, taking all L-algebras in the Grothendieck universe $\mathcal{U}_1$, we get a category $\mathcal{S}_1$ which is (obviously) monadic over $\mathtt{SET}$, the category of $\mathcal{U}_1$-sets, and thus the factorization fact will hold in $\mathcal{S}_1$. Note that, starting with an arrow in the subcategory $\mathcal{S}$ of $\mathcal{S}_1$, the construction of the regular factorization will stay entirely within $\mathcal{S}$.]

Let us hasten to point out that there is a completely elementary (even trivial) proof of the Beth definability theorem for Boolean algebras; nevertheless, the above proof is suggestive of what is to come later.

2. FACTORIZATIONS IN 2-CATEGORIES

The basic notions around 2-categories that are explained in [M/P] will suffice for the exposition that follows. Most everything we say in this section in a general way about 2-categories could be repeated, essentially without change, for bicategories (see [Be], [S]). This section may also be omitted; the next section, and the end of section 12, contain alternative (less conceptual) formulations of the theorems and proofs.

The role $\mathtt{Set}_{\mathcal{U}}$, the category of $\mathcal{U}$-sets with $\mathcal{U}$ a Grothendieck universe, plays among categories is taken up by $\mathit{Cat}_{\mathcal{U}}$, the 2-category of $\mathcal{U}$-categories, functors and natural transformations. We write $\mathcal{CAT}$ for $\mathit{Cat}_{\mathcal{U}_1}$; thus, $\mathtt{Set}$ is an object in $\mathcal{CAT}$, among others.

Perhaps the most natural counterpart in $\mathcal{CAT}$ of the surjective/injective factorization in $\mathtt{Set}$ is the *essentially-surjective-on objects/full-and-faithful* (abbreviated e.s./f.f.) factorization. Every functor $F:\mathbf{A}\longrightarrow\mathbf{B}$ may be factored in the form of the diagram

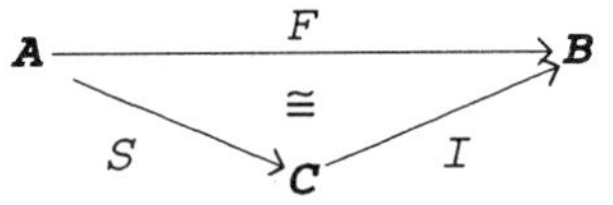

commutative up to isomorphism, where S is essentially surjective on objects (meaning that for all $B\in\mathbf{B}$ there is $A\in\mathbf{A}$ such that $SA\cong B$), and I is full and faithful (meaning that I induces bijections $\mathrm{Hom}_{\mathbf{A}}(A, A')\longrightarrow\mathrm{Hom}_{\mathbf{B}}(IA, IA')$). Moreover, such a factorization is unique up to equivalence, in a precise sense that is easy to guess.

The *(2-)regular factorization* takes place in any 2-category (or even bicategory) which is sufficiently bicomplete and bicocomplete. The regular factorization is defined up to isomorphism; the 2-regular one up to equivalence, in agreement with the fact that the universal properties (bilimits, bicolimits) determine the relevant diagrams up to equivalence only. It can be recovered as the natural answer to the question: "what bicategorical universal construction generalizing the regular factorization will give us the e.s./f.f. factorization in $\mathcal{CAT}$?" The 2-regular factorization is closely related to the construction of the category of *descent data* corresponding to an arrow in the base-category of a fibration.

The general concept of the 2-regular factorization is due to R. Street, although he does not use the same terminology; see [S2] and [S3]. In fact, Street introduces the notions of regular 2-category (bicategory) and exact 2-category (bicategory); the 2-categorical version is the subject of [S2], the bicategorical one that of [S3]. $\mathcal{CAT}$ is the prototype of an exact 2-category (and also, of an exact bicategory). The 2-regular factorization plays the same role in the theory of exact 2-(bi-)categories as the ordinary regular factorization plays in the theory of exact categories (for the latter, see [B]). As a matter of fact, we do not need anything but the definition of the 2-regular factorization, which we will give in a terminology slightly different from that of [S2], [S3].

Professor Street has asked me to communicate a correction to his two papers mentioned above. The theorem at the bottom of page 284 in [S3], and Theorem 1.22 (p. 257) in [S2] are false. The correction is that the conclusions of the theorems have to be taken as further axioms for a regular 2-category/bicategory.

In what follows, we put ourselves in a fixed 2-category and define the notion of 2-regular factorization in that 2-category (we use the *bicategorical* version in the context of *2-categories*).

A *c-complex* is given by data as follows: objects and arrows as in the diagram

$$\begin{array}{ccccc} & \xrightarrow{p_{01}} & & \xrightarrow{p_0} & \\ E & \xrightarrow{p_{02}} & K & \xleftarrow{\Delta} & A \; ; \\ & \xrightarrow{p_{12}} & & \xrightarrow{p_1} & \end{array} \tag{1}$$

and 2-arrows that are isomorphisms as in

$$\iota_0 : p_0\Delta \cong 1_A \; , \; \iota_1 : p_1\Delta \cong 1_A$$

$$\rho_1 : p_1 p_{01} \cong p_0 p_{12} \; , \; \rho_0 : p_0 p_{01} \cong p_0 p_{02} \; , \; \rho_2 : p_1 p_{12} \cong p_1 p_{02} \; .$$

Suppose we have a c-complex as above, an arrow

$$F : A \longrightarrow B$$

and a 2-arrow (not necessarily an isomorphism)

$$\mu \; : \; FP_0 \longrightarrow FP_1 \; .$$

The arrow $\tilde{1}_F : FP_0\Delta \longrightarrow FP_1\Delta$ is defined to make

$$\begin{array}{ccc} FP_0\Delta & \xrightarrow{F\iota_0} & F \\ {\scriptstyle \tilde{1}_F}\downarrow & & \downarrow{\scriptstyle 1_F} \\ FP_1\Delta & \xrightarrow[F\iota_1]{} & F \end{array}$$

commute. The 2-arrow

$$\mu P_{12} \,\tilde{\circ}\, \mu P_{01} \;\; : \;\; FP_0 P_{02} \longrightarrow FP_1 P_{02}$$

is defined as the one that makes the hexagon

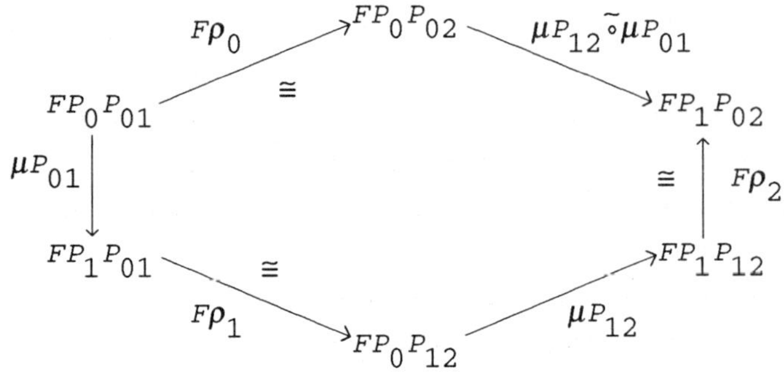

commute. We say that (F, μ) *equalizes* the given c-complex if the equalities

$$\mu\Delta = \tilde{1}_F$$

$$\mu P_{02} = \mu P_{12} \tilde{\circ} \mu P_{01}$$

hold.

Given an arrow $F: A \longrightarrow B$, we may ask for the c-complex $\mathcal{C}$ (in the above notation) together with $\mu: FP_0 \longrightarrow FP_1$ so that (F, μ) equalizes $\mathcal{C}$ and $(\mathcal{C},\mu)$ is universal with this property; the resulting pair $(\mathcal{C},\mu)$ is called the 2-*kernel complex of* F (sometimes, $\mathcal{C}$ by itself is referred to as the 2-kernel complex of F). Instead of spelling out the universal property (a limit-like one), we build up the 2-kernel pair using more familiar pieces.

First we get μ to make

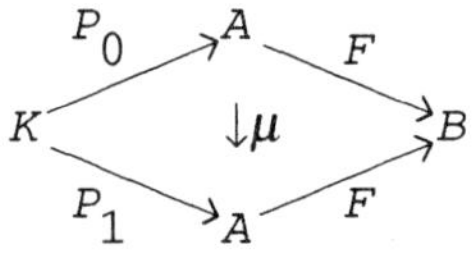

(2)

a *comma square*. To explain the universal property of the comma square, let, besides $F: A \to B$, also the object D be (temporarily) fixed. The category $\mathrm{Comma}_F(D)$ has objects triples (Q_0, Q_1, ν) as in

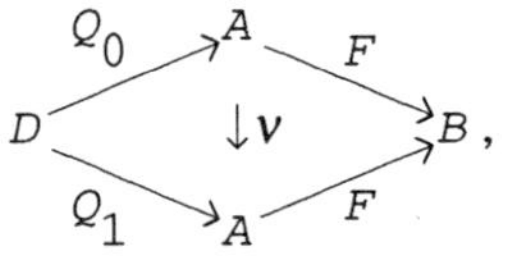

and arrows $(Q_0, Q_1, \nu) \longrightarrow (Q_0', Q_1', \nu')$ pairs $(\zeta_0: Q_0 \to Q_0', \zeta_1: Q_1 \to Q_1')$ making

$$\begin{array}{ccc} FQ_0 & \xrightarrow{\zeta_0} & FQ_0' \\ {\scriptstyle \nu}\downarrow & & \downarrow{\scriptstyle \nu'} \\ FQ_1 & \xrightarrow[\zeta_1]{} & FQ_1' \end{array}$$

commute; the composition is the obvious one. With any $\mathcal{P} = (P_0, P_1, \mu) \in \mathrm{Comma}_F(K)$, and $G: D \to K$, we have

$$\mathcal{P}^*(G) \underset{\mathrm{def}}{=} (P_0 G, P_1 G, \mu G) \in \mathrm{Comma}_F(D) ;$$

and if $\gamma: G \to H$ is an arrow in $\mathrm{Comma}_F(D)$, then

$$\mathcal{P}^*(\gamma) \underset{\mathrm{def}}{=} (P_0\gamma, P_1\gamma) : \mathcal{P}^*(G) \longrightarrow \mathcal{P}^*(H) ;$$

these specifications define a functor $\mathcal{P}^* : \mathrm{Hom}(D, K) \longrightarrow \mathrm{Comma}_F(D)$. The square $\mathcal{P}$ as in (1) is a *comma square* if for every D, the functor $\mathcal{P}^*$ is an *equivalence* of categories.

In case of $\mathcal{CAT}$, C in the comma square above can be taken to be the comma-category F/F with objects $(a, b, Fa \longrightarrow Fb)$ (hence the name for the notion). Moreover, in this case, in the description of the universal property, $\mathcal{P}^*$ will be an isomorphism of categories.

Secondly, we form the *bipullback*

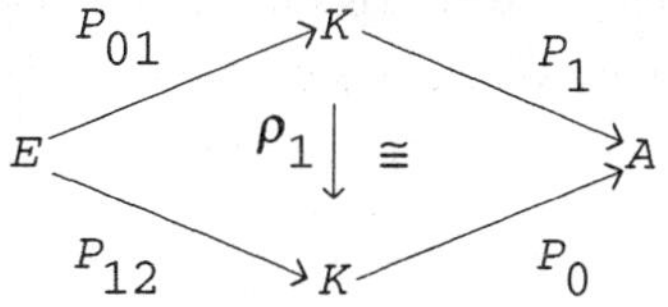

which has a universal property analogous to the one explained above, except that now ρ_1 is restricted to be an isomorphism. If we are in $\mathcal{CAT}$, E may be taken to be the category whose objects are entities of the form $(a, b, c, Fa \xrightarrow{\alpha} Fb \xrightarrow{\beta} Fc)$; P_{01} maps the latter to $(a, b, Fa \xrightarrow{\alpha} Fb)$, P_{12} to $(b, c, Fb \xrightarrow{\beta} Fc)$, and ρ_1 is the identity.

Thirdly, using the universal property of the comma square (2), and consulting the diagram

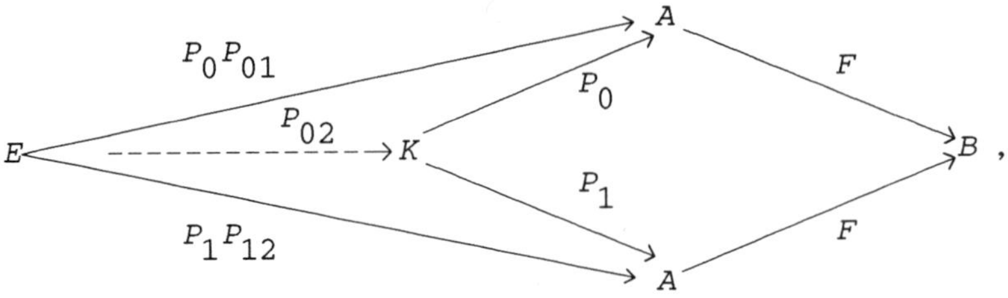

we choose P_{02} with the other two isomorphisms ρ_0 , ρ_2 , satisfying the second identity in the definition of "c-complex". In fact, we have that

$$(P_0P_{01},\ P_1P_{12},\ \mu P_{12}\circ F\rho_1\circ\mu P_{01}) \in \mathrm{Comma}_F(E)\ ,$$

and we use that $\mathcal{P}^*:\mathrm{Hom}(E,\ K) \longrightarrow \mathrm{Comma}_F(E)$ is essentially surjective on objects. Δ , together with ι_0 , ι_1 , is also obtained by using the universal property of the comma square.

This completes the definition of the 2-kernel complex of an arrow in a 2-category. The definition is rather formidable as it stands. In many important cases, it can be simplified by choosing all the 2-isomorphisms ι_0 , ι_1 , ρ_0 , ρ_1 , ρ_2 in it to be identities; this is the case with $\mathcal{CAT}$. If so, $1_{\tilde{F}} = 1_F$, and $\mu P_{12}\tilde{\circ}\mu P_{01} = \mu P_{12}\circ\mu P_{01}$, in particular. Further note that, in $\mathcal{CAT}$, with the choice of E as said above, P_{02} maps the object $(a,\ b,\ c,\ Fa\xrightarrow{\alpha}Fb\xrightarrow{\beta}Fc)$ to $(a,\ c,\ Fa\xrightarrow{\beta\circ\alpha}Fc)$.

Given *any* c-complex $\mathcal{C}$ in the notation above , the *quotient of* $\mathcal{C}$ is a pair $(S{:}A\to C,\ \delta)$ equalizing $\mathcal{C}$, and universal with this property. This means the following. With C any object, $\mathrm{Equ}_{\mathcal{C}}(C)$ is, by definition, the category whose objects are the pairs $(H{:}A\to C,\ \mu)$ equalizing $\mathcal{C}$, and whose arrows $(H{:}A\to C,\ \mu) \to (H'{:}A\to C,\ \mu')$ are $\varphi{:}H\to H'$ such that

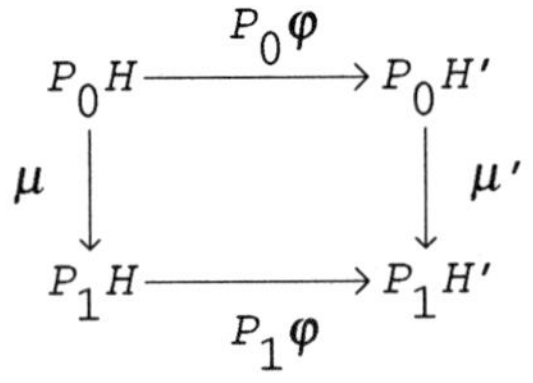

commutes. With $q = (S{:}A\longrightarrow C,\ \delta)$ and any object D we have the functor $q_D^*{:}\mathrm{Hom}(C,\ D) \to \mathrm{Equ}_{\mathcal{C}}(D)$ defined in the obvious way. $(S,\ \delta)$ is a *quotient of* $\mathcal{C}$ if for all D , q_D^* is an equivalence of categories.

Starting with an arrow $F{:}A\longrightarrow B$, we let $(S{:}A\longrightarrow C,\ \delta)$ be the quotient of its 2-kernel complex $(\mathcal{C},\mu)$, with $\mathcal{C}$ in the notation of (1). By the universal property of $(S,\ \delta)$, we obtain an arrow $I{:}C\longrightarrow B$ such that

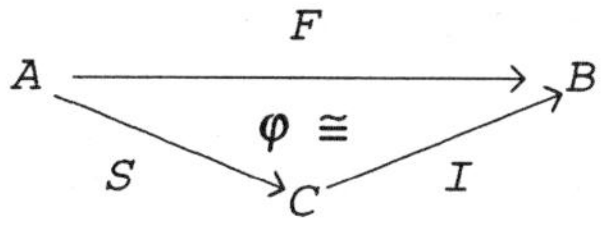

(3)

commutes up to an isomorphism called φ for which

$$
\begin{array}{ccc}
P_0 F & \xrightarrow{\ P_0\varphi\ } & P_0 IS \\[4pt]
\mu \downarrow & & \downarrow I\delta \\[4pt]
P_1 F & \xrightarrow[\ P_1\varphi\ ]{} & P_1 IS
\end{array}
$$

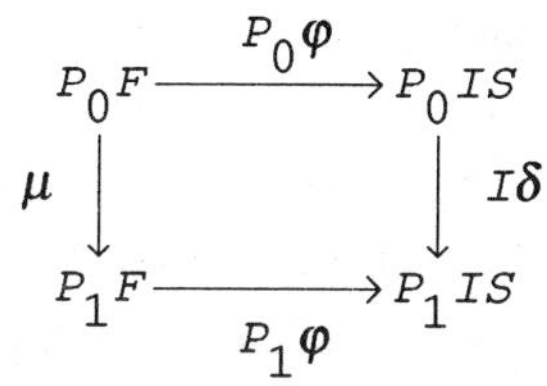

commutes. Any diagram (3) obtained in this way is a 2-*regular factorization of* F ; as we said above, the 2-regular factorization is determined up to equivalence.

To repeat the first motivation for this definition, the regular factorization of an arrow (functor) F in $\mathcal{CAT}$, S is essentially surjective on objects, I is full and faithful. In fact, in calculating the quotient $(S : A \to C,\ \delta : SP_0 \to SP_1)$ of the 2-kernel pair of F (the latter calculated as explained above), we may take C to be the category whose objects are the objects of A , and whose morphisms $a \dashrightarrow b$ are the arrows $Fa \longrightarrow Fb$ in B , with composition specified in the natural way; S maps a to a (thus, S is a *bijection* on objects), an arrow $\alpha : a \to b$ to $F\alpha$, and φ is taken to be the identity. Moreover, $\delta_{(a,\ b,\ f : Fa \to Fb)} = f$. The functor $(q_D^{*})^{-1}$ will take any $(H : A \to D, \mu) \in \mathrm{Equ}_C(D)$ into the functor $Z : Q \to D$ for which $Z(a) = H(a)$ and $Z([f : Fa \to Fb] : a \dashrightarrow b) = \mu_{(a,\ b,\ f)}$. The most interesting part of the verification is to see that $Z \circ S = H$.

The *opposite* $\mathcal{A}^{\mathrm{op}}$ of a 2-category $\mathcal{A}$ has the same objects, arrows and 2-arrows as $\mathcal{A}$, except that all arrows are reversed: what was $F : A \to B$ in $\mathcal{A}$ is $F : B \to A$ in $\mathcal{A}^{\mathrm{op}}$; the 2-arrows do not change direction. We talk about the (2-)*coregular factorization* of an arrow in $\mathcal{A}$ meaning the regular factorization of the corresponding arrow in $\mathcal{A}^{\mathrm{op}}$; note that limits $\mathcal{A}^{\mathrm{op}}$ become colimits in $\mathcal{A}$, and vice versa.

When $\mathcal{A}$ is a concrete 2-category (with objects structured categories), the object C constructed in the coregular factorization of $F : A \to B$ in $\mathcal{A}$ is called the *category of descent data of* F , and is denoted $\mathrm{Des}(F)$. In fact, this terminology is close to that of descent theory (see [G], [J/T]) when all 2-arrows of $\mathcal{A}$ are isomorphisms; in the general case, we might want to talk about "lax descent" instead; in this paper, we are interested in proper descent, inasmuch our 2-arrows will be isomorphisms.

Note that when we apply the regular factorization in a 2-category in which all 2-cells are isomorphisms (that is, in a groupoid-enriched category), then the comma object becomes a bipullback; thus, the kernel complex in this case is constructed as two consecutive (bi)pullbacks, and additional resulting morphisms.

Unlike the case with ordinary concrete categories, it happens more rarely that a concrete 2-category has the described property of $\mathcal{CAT}$, namely that the regular factorization in it coincides with the e.s./f.f. factorization; let us call this property, at least temporarily, *(2-)pre-regularity*. Certainly, monadicity over $\mathcal{CAT}$ is not enough for pre-regularity. E.g., the 2-category $\mathcal{LEX}$ of categories with finite limits and finite-limit-preserving functors is monadic over $\mathcal{CAT}$ (see [B/K/P]), but arrows in it do not, in general, have an e.s./f.f. factorization at all. Even in cases when all arrows can be obviously e.s./f.f.-factored, the question whether the factorization can be obtained as the regular one may be a further problem; in fact, it is in the situation we will have to consider later in this paper.

The problem with $\mathcal{LEX}$ in this respect lies in the fact that the operation of equalizer creates a new object on the basis of arrows as arguments. "Doctrines" where we do not have this "problem" such as that of cartesian (finite-product) categories, cartesian closed categories (with isomorphisms as 2-cells), etc, pre-regularity does hold.

I do not see a useful general way of saying when pre-regularity holds; in our story, it will be an important element that in a particular 2-category (of "ultracategories") the property holds at least sufficiently often.

The 2-regular factorization is closely related to descent theory. For instance, the main result of A. Joyal and M. Tierney in [J/T], the descent theorem for open geometric morphisms can be stated by saying that

the 2-regular factorization of an open geometric morphism in $\mathtt{Top}^{\mathtt{iso}}$ *, the 2-category of (Grothendieck) toposes with 2-cells restricted to isomorphisms, is the surjection/inclusion factorization.*

For the concept used here, see [J] and [J/T]. The usual statement is obtained when the regular factorization is applied to an open geometric morphism which is already a surjection; in this case, we get the geometric morphism itself as the "surjective" part of the regular factorization; this is equivalent to saying that the geometric morphism is an effective descent morphism in the terminology of [J/T].

We will be interested in the 2-categories $\mathcal{PRETOP}$, and $\mathcal{BOOLEPRETOP}^*$ (or $\mathcal{BP}^*$, more briefly). The first has objects the pretoposes in $\mathcal{U}_1$, the pretopos functors between them (those preserving the pretopos structure) as arrows, and all natural transformations between the latter as 2-arrows. For the concept of pretopos, see [M/R] , or [M3] . $\mathcal{BP}^*$ has objects the Boolean pretoposes, that is, those in which every subobject lattice is complemented (is a Boolean algebra; see also [M3]), whose arrows are pretopos functors (these preserve Boolean complements automatically), and whose 2-arrows are the *isomorphisms* between the latter. $\mathcal{BP}^*$

may be called the *groupoid-enriched* category of Boolean pretoposes because the hom's in $\mathcal{BP}^*$ are groupoids, that is, categories with all arrows isomorphisms.

The main reason to restrict attention to isomorphism 2-arrows is that this is the way to get to our main aim, the descent theorem, which is about isomorphisms between models. Another reason is that in the 2-category $\mathcal{BP}$, with all natural transformations retained, is not a "good" category; e.g., the cocomma square (the dual of "comma-square") does not in general exist, and thus, the coregular factorization cannot always be carried out!

Another factorization is related to *quotient morphisms* and *conservative morphisms.*

A *concrete* 2-category is a 2-category with a forgetful (2-)functor to $\mathcal{CAT}$. Usually we require of the forgetful functor that it reflect isomorphism 2-cells and equivalence 1-cells (if a 2-cell in the category is mapped by the forgetful functor into an isomorphism, then it is an isomorphism itself; if a 1-cell is mapped into an equivalence, it is an equivalence itself). When we say that an arrow in a concrete 2-category is e.s., or f.f., we mean that its image under the forgetful functor has the corresponding property. We use the absolute-value symbol $|\ |$ to denote the effect of the forgetful functor, and call $|A|$ the *underlying category* (*functor*) of A.

In any concrete 2-category, we can talk about the quotient/conservative (q./c.) factorization; and in many concrete 2-categories, this factorization exists and has the expected properties.

Let us place the discussion into a fixed concrete 2-category. Given any object and a class Σ of arrows in the underlying category of A, $Q:A\longrightarrow A[\Sigma^{-1}]$ is defined as the universal morphism $A\longrightarrow B$ so that every $\sigma\in\Sigma$ becomes an isomorphism in $|B|$. In other words, $|Q|(\sigma)$ is an isomorphism for all $\sigma\in\Sigma$, and for any object B in our 2-category, the functor

$$Q^* : \mathrm{Hom}(A[\Sigma^{-1}], B) \longrightarrow \mathrm{Hom}_{\Sigma^{-1}}(A, B)$$

defined by composition, with the codomain denoting the full subcategory of $\mathrm{Hom}(A, B)$ of those morphisms G for which $|G|(\sigma)$ is an isomorphism for all $\sigma\in\Sigma$, is an equivalence of categories. A morphism $F:A\longrightarrow B$ is a *quotient* morphism if F has the universal property of $Q:A\longrightarrow A[\Sigma^{-1}]$, with Σ (necessarily) the set Inv_F of those arrows whose F-image is invertible in B. A morphism $F:A\longrightarrow B$ is *conservative* if its underlying functor $|F|$ reflects isomorphisms: if $|F|(\alpha)$ is an isomorphism, so is α.

The construction of $A[\Sigma^{-1}]$ is a colimit-like construction, and it can be performed in all "reasonable" 2-categories.

Now, if $F:A\longrightarrow B$ is any morphism, then, using the universal property of $A[\mathrm{Inv}_F^{-1}]$, we certainly have a factorization

$$\begin{array}{ccc} A & \xrightarrow{\quad F \quad} & B \\ & \cong & \\ & {\scriptstyle Q} \searrow \quad A[\mathrm{Inv}_F^{-1}] \quad \nearrow {\scriptstyle J} & \end{array}$$

whose left leg is, by definition, a quotient. In good cases, the right leg is conservative (it is certainly conservative with respect to all $A[\mathrm{Inv}_F^{-1}]$-arrows coming from A, but what about other arrows ?). We say that an arrow F in our 2-category *has a q./c. factorization* if J is conservative; the 2-category has the factorization if all arrows in it do. It is easy to see that the q./c. factorization (if exists) is unique up to equivalence; in particular, a morphism which is both a quotient and conservative, is an equivalence.

The two 2-categories $\mathcal{PRETOP}$ and $\mathcal{BP}^*$ mentioned above are "good"; for instance, they are bicomplete and bicocomplete, ensuring that all the universal constructions used above can be carried out in them. In [M3], it is proved that both $\mathcal{PRETOP}$ and $\mathcal{BP}^*$ have the q./c. factorization (a fact belonging to the folklore). Moreover, [M3] shows another well-known characterization of quotient morphisms in these 2-categories that we now describe.

Let $I: S \longrightarrow T$ be a pretopos functor between pretoposes, $X \in T$. I *covers* X if there are $A \in S$ and a (regular) epimorphism $IA \twoheadrightarrow X$ in T. I *finitely subcovers* X if there are $n < \omega$, objects A_i $(i<n)$ in S, monomorphisms $Y_i \rightarrowtail IA_i$ $(i<n)$ in T, and an epimorphism $\coprod_{i<n} Y_i \twoheadrightarrow X$. (The intermediate concepts "subcovers" , and "finitely covers" have the expected descriptions). I *covers* or *finitely subcovers* T if it does all objects in T.

We say that I is *full on subobjects* if for any $A \in S$, the induced mapping (a lattice homomorphism) $\mathrm{Sub}_S(A) \longrightarrow \mathrm{Sub}_T(IA)$ is surjective.

1. Proposition. ([M3], 2.4.7) A morphism in $\mathcal{PRETOP}$ or in $\mathcal{BP}^*$ is a quotient morphism iff it is full on subobjects and it covers its codomain, and also iff it is full on subobjects and finitely subcovers its codomain.

In the correspondence between theories and pretoposes (see [M/R]), a quotient morphism $I: S \longrightarrow T$ of pretoposes corresponds to an extension of a theory S by adding only new axioms, but no new extra-logical symbols. (The reason that a quotient morphism is not necessarily essentially surjective on objects is that adding new axioms may create new definable equivalence relations.) A conservative morphism corresponds to a conservative interpretation of theories when in the extension theory no new theorems in the language of the extended theory are provable.

Marek Zawadowski's lax descent theorem [Z1], [Z2] on pretoposes says that

the coregular factorization of a morphism of pretoposes (*in the 2-category of pretoposes*) *is the quotient/conservative factorization.*

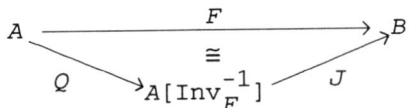

Zawadowski also showed that

the same conclusion holds for the 2-category of Barr-exact categories

(for definitions, see e.g. [B] or [M5]).

It may be mentioned that the Joyal/Tierney descent theorem can also be stated as the coincidence of the coregular factorization with the quotient/conservative one, for the left-adjoint part of open geometric morphisms, in the 2-category opposite to $\mathtt{Top}^{\mathtt{iso}}$.

The main new result to be proved in this paper is

2. Theorem. The coregular factorization in $\mathcal{BP}^*$ is the quotient/conservative factorization.

Actually, we will deal with $\mathcal{B}p^* = \mathcal{B}oole\mathcal{P}retop^*$, the 2-category of small Boolean pretoposes (with isomorphims 2-cells), rather than the more comprehensive $\mathcal{BOOLEPRETOP}^* = \mathcal{BP}^*$. However, in the proof of the result nothing particular is used on the universe $\mathcal{U}_0$ we chose to specify $\mathcal{B}p^* = \mathcal{B}p^*_{\mathcal{U}_0}$; changing $\mathcal{U}_0$ to $\mathcal{U}_1$ gives the result for $\mathcal{BP}^* = \mathcal{B}p^*_{\mathcal{U}_1}$.

We turn to stating the 2-categorical version of the Zawadowski setup on which our plan of the proof of the last theorem is based. Some preliminaries are needed for this.

Our 2-categories, unlike the categories in Section 1, will contain objects that are not necessarily small; at the same time, the small objects will have a distinguished role. An object of a concrete 2-category is called *small* if its underlying category is essentially small, that is, it is equivalent to a small category.

Following Max Kelly's advice, we drop superfluous qualifiers such as "2-" whenever reasonable; e.g., a "functor" between 2-categories is necessarily a 2-functor; if we wanted to talk about the ordinary functor between the underlying categories, we would do so explicitly.

We will speak about *inclusions* in one of the concrete 2-categories we deal with. In the application we have in mind for Boolean pretoposes, the underlying functors of inclusions of so-called ultragroupoids will be, in particular, full and faithful, but the converse will not hold. The reason is that the structure of ultragroupoids will involve *relations* on arrows; an inclusion will also have to *reflect* the relations. In the case of the Zawadowski setup for pretoposes, inclusions will coincide with arrows with full and faithful underlying functors. The e.s./incl. factorization is one in which the first factor is essentially surjective on objects, the second is an inclusion.

The Zawadowski setup consists, by definition, of two concrete 2-categories $\mathcal{P}$ and $\mathcal{U}$, and a (strict) (2-)adjunction

$$\mathcal{P}^{op} \underset{G}{\overset{F}{\rightleftarrows}} \mathcal{U} \quad , \quad F \dashv G \text{, with counit } \varepsilon ,$$

together with a class of arrows in $\mathcal{U}$ called *inclusions*, satisfying the following conditions:

Z0. $\mathcal{P}$ is finitely bicomplete and bicocomplete, and the forgetful functor on $\mathcal{P}$ preserves finite bilimits.

Z1. For every small $A \in \mathcal{P}$, $\varepsilon_A : A \longrightarrow FGA$ (a morphism in $\mathcal{P}$, rather than $\mathcal{P}^{op}$) is conservative.

Z2. If $A \in \mathcal{P}$ is small, and the morphism $\Phi : X \longrightarrow GA$ in $\mathcal{U}$ is an inclusion in $\mathcal{U}$, then the transpose of Φ, $\Phi^{\#} : A \longrightarrow FX$, is a quotient morphism in $\mathcal{P}$.

Z3. For any arrow I between small objects in $\mathcal{P}$, both the e.s./incl. factorization and the (2-)regular factorization of $G(I)$ exist in $\mathcal{U}$, and they coincide.

3. Proposition. In any concrete 2-category that takes part in a Zawadowski setup "on the left", the 2-coregular factorization of any morphism between small objects is the quotient/conservative factorization.

Proof. The proof follows the proof of 1.1. closely; we make a few remarks only. As a left 2-adjoint, F preserves bicolimits, therefore it takes a quotient of a c-complex in $\mathcal{U}$ into a "coquotient" of the corresponding "cocomplex" in $\mathcal{P}$. Similarly, G takes the 2-cokernel complex of an arrow in $\mathcal{P}$ into a 2-kernel complex of the corresponding arrow in $\mathcal{U}$. [] 3

Zawadowski's specific theorem is, in essence, the statement that

$\mathcal{PRETOP}$, the 2-category of pretoposes takes part in a Zawadowski setup on the left;

and also,

the same holds for Barr-exact categories in place of pretoposes.

Zawadowski's descent theorems follow. (For readers of [Z2], let us note that Zawadowski does not in fact follow the precise schema of what we call the "Zawadowski setup"; thus, our version here is an interpretation, rather than a literal copy, of the basic outline of Zawadowski's proof.)

We have

4. Theorem. $\mathcal{BP}^*$, the groupoid-enriched category of Boolean pretoposes takes part in a Zawadowski setup on the left.

2. follows from 3. and 4.; the latter's proof will occupy us for the rest of the paper.

3. DEFINABLE FUNCTORS

A *theory* $\mathcal{T}$ is a pair (L, T), with L a possibly many-sorted language for first order logic, and T a set of sentences in full first order logic over L. $\mathrm{Mod}(\mathcal{T})$ is the category whose objects are the L-structures M (with possibly empty domains $M(A)$, for sorts A in L) that are models of T in the ordinary sense, and whose morphisms are the elementary embeddings. $\mathrm{Mod}^*(\mathcal{T})$ is the subcategory of $\mathrm{Mod}(\mathcal{T})$ with the same objects as $\mathrm{Mod}(\mathcal{T})$, and with the isomorphisms in $\mathrm{Mod}(\mathcal{T})$ as arrows. We will be mainly interested in the groupoid $\mathrm{Mod}^*(\mathcal{T})$, rather than the category $\mathrm{Mod}(\mathcal{T})$.

Each (first order) L-formula $\varphi(\vec{x})$ gives rise to a functor

$$[\vec{x}:\varphi] : \mathrm{Mod}(\mathcal{T}) \longrightarrow \mathrm{Set},$$

which maps M to the set $M[\vec{x}:\varphi] \overset{=}{_{\mathrm{def}}} \{\vec{a}: M \models \varphi[\vec{a}/\vec{x}]\}$, the interpretation of φ in M, and maps elementary maps to the induced restriction-functions. Let us call any functor $\mathrm{Mod}(\mathcal{T}) \longrightarrow \mathrm{Set}$ so obtained *definable.*

The restriction to $\mathrm{Mod}^*(\mathcal{T})$ of a definable functor on $\mathrm{Mod}(\mathcal{T})$ will also be called *definable*.

Consider now a theory $\mathcal{T} = (L', T)$, and a sublanguage L of L'. Let $\mathcal{S} = (L, S)$ be the theory whose axioms (elements of S) are the L-consequences of T. Consider the category $\mathrm{Mod}(\mathcal{T}/L)$ whose objects are the L'-models of T, but whose arrows are L-elementary maps: an arrow $M \overset{\cdot}{\longrightarrow} N$ in $\mathrm{Mod}(\mathcal{T}/L)$ is an elementary map $M{\restriction}L \longrightarrow N{\restriction}L$ between the reducts. We have the forgetful functor $F: \mathrm{Mod}(\mathcal{T}/L) \rightarrow \mathrm{Mod}(\mathcal{S})$ for which $F(M) = M{\restriction}L$ and F acts (essentially) as the identity on arrows; F is full and faithful. Thus, up to equivalence of categories, $\mathrm{Mod}(\mathcal{T}/L)$ is the same as the full subcategory of $\mathrm{Mod}(\mathcal{S})$ whose models can be expanded to $\mathcal{T}$-models. In A. Tarski's terminology, the class of objects of $\mathrm{Mod}(\mathcal{T}/L)$ is a typical PC_Δ-class; thus, a category of the form $\mathrm{Mod}(\mathcal{T}/L)$ might be called a PC_Δ-category. $\mathrm{Mod}^*(\mathcal{T}/L)$ has the same objects as $\mathrm{Mod}(\mathcal{T}/L)$, and the isomorphisms of $\mathrm{Mod}(\mathcal{T}/L)$ (the isomorphisms between the corresponding reducts) as arrows.

Now, here is a concept of "definable functor" of the form $F: \mathrm{Mod}^*(\mathcal{T}/L) \longrightarrow \mathrm{Set}^*$, intended as the "widest reasonable" such notion. Let us take a formula $\varphi(\vec{x})$ over L'; φ will define the effect of F on objects; $F(M) = M[\vec{x}:\varphi] =$

$\varphi[M]$. How are we to *define* the effect of the functor on an arrow $f:M\stackrel{\cdot}{\longrightarrow}N$ (i.e., $f:M\restriction L\xrightarrow{\cong}N\restriction L$) ? Well, we can consider the new theory $\mathcal{T}+_L\mathcal{T}$ that talks about two L'-structures, and an isomorphism map between their L-reducts. This is obtained by taking the language L^+ which is the union of two disjoint copies of L' , and which has, in addition, an operation symbol $\mathbf{f}_A:A_{(0)}\longrightarrow A_{(1)}$ for each sort A of L ; $A_{(0)}$, $A_{(1)}$ are the two copies of A in L^+ . The axioms T^+ of $\mathcal{T}+_L\mathcal{T}$ are the two copies of all the axioms of T , and the sentences expressing that the system of the $\mathbf{f}_A$ form an isomorphism between the L-reducts of the two "parts", each an L'-model of T, of a model. $\mathcal{T}+_L\mathcal{T}\ \overline{\overline{\mathrm{def}}}\ (L^+, T^+)$. Thus, a model of $\mathcal{T}+_L\mathcal{T}$ is a composite structure of the form (M, N, f) , where M and N are models of $\mathcal{T}$, and f is an (L-)isomorphism $f:M\restriction L\xrightarrow{\cong}N\restriction L$.

Now, given the L'-formula $\varphi(\vec{x})$ to define the object-effect of our functor F , we may consider the two copies $\varphi_{(0)}$, $\varphi_{(1)}$; these are formulas in $\mathcal{T}+_L\mathcal{T}$ (of L^+) obtained in the obvious way. Let $\mu(\vec{x}, \vec{y})$ be a formula of L^+ for which it is provable in $\mathcal{T}+_L\mathcal{T}$ that it defines a bijection from the set defined by $\varphi_{(0)}$ to that by $\varphi_{(1)}$:

$$T^+\vdash\forall\vec{x}\vec{y}(\mu(\vec{x},\vec{y})\longrightarrow\varphi_{(0)}(\vec{x})\wedge\varphi_{(1)}(\vec{y}))$$
$$\wedge\ \forall\vec{x}(\varphi_{(0)}(\vec{x})\longrightarrow\exists!\vec{y}\mu(\vec{x},\vec{y}))\ \wedge\ \forall\vec{y}(\varphi_{(1)}(\vec{y})\longrightarrow\exists!\vec{x}\mu(\vec{x},\vec{y}))\ .$$

In other words, for any model (M, N, f) of $\mathcal{T}+_L\mathcal{T}$, the interpretation $\mu^{(M,N,f)}$ of μ , a relation $\mu^{(M,N,f)}\subset\prod_i M(X_i)\times\prod_i N(X_i)$ where the X_i are the respective sorts of the variables in $\vec{x}$ (and $\vec{y}$) , defines a bijection, also denoted by

$$\mu^{(M,N,f)}:\varphi^M\xrightarrow{\cong}\varphi^N,$$

with φ^M , φ^N the interpretations of φ in M and N, resp.

Let us call a pair (φ,μ) of formulas, φ in $\mathcal{T}$, μ in $\mathcal{T}+_L\mathcal{T}$, satisfying the conditions described above a *potential isomorphism functor specification* (pifs). Any pifs (φ,μ) defines, via the assignments

$$M\longmapsto\varphi^M \qquad (\ M\in\mathrm{Mod}^*(\mathcal{T}/L)\ )$$

$$f\longmapsto\mu^{(M,N,f)} \qquad (\ f:M\stackrel{\cdot}{\longrightarrow}N \text{ in } \mathrm{Mod}^*(\mathcal{T}/L)\ ),$$

a morphism $[\varphi,\mu]$ of graphs from the underlying graph of $\mathrm{Mod}^*(\mathcal{T}/L)$ to the underlying graph of Set^* . It is an *isomorphism functor specification* (ifs), or more specifically, a $\mathcal{T}/L$*-ifs* if $[\varphi, \mu]$ is a *functor*

$$[\varphi, \mu] : \mathrm{Mod}^*(\mathcal{T}/L) \longrightarrow \mathrm{Set}^*,$$

that is, if it takes identity arrows to identity arrows, and preserves composition. Expressed directly, this means that

$$\mu^{(M, M, 1_{M\restriction L})} = 1_{\varphi^M},$$

and

$$\mu^{(N, P, g)} \circ \mu^{(M, N, f)} = \mu^{(M, P, f\circ g)}$$

for all models M of $\mathcal{T}$, and models (M, N, f), (N, P, g) of $\mathcal{T}+_L\mathcal{T}$.

Note that, by the use of a certain straightforward theory $\mathcal{T}+_L\mathcal{T}+_L\mathcal{T}$ (see Section 9), the conditions for a pair (φ,μ) to be an ifs may be formulated as the provability of certain sentences from specific axioms. As a consequence, if we take our theory $\mathcal{T}$ to be countable, and have it Gödel-numbered in the standard way, then the condition for (φ,μ) to be a $\mathcal{T}/L$-ifs is Σ_1^0 in the real coding (L', T, L).

A *definable functor* $\mathrm{Mod}^*(\mathcal{T}/L) \longrightarrow \mathrm{Set}^*$ is, by definition, one of the form $[\varphi,\mu]:\mathrm{Mod}^*(\mathcal{T}/L) \longrightarrow \mathrm{Set}^*$, with an ifs (φ,μ).

The definition of a definable functor on a PC_Δ-category does not tell us how to *obtain* such a functor, unlike the definition of definable functor on an elementary category (one of the form $\mathrm{Mod}(\mathcal{T})$, or $\mathrm{Mod}^*(\mathcal{T})$) which tells us that any formula of the underlying language gives rise to such a functor; in the PC_Δ-case, there are *conditions* on the formulas that are non-trivial to satisfy.

One obvious way to get a definable functor on $\mathrm{Mod}(\mathcal{T}/L)$ is take any L-formula (!) $\varphi(\vec{x})$, and *restrict* the definable functor $[\vec{x}:\varphi]:\mathrm{Mod}(\mathcal{S}) \longrightarrow \mathrm{Set}$, to $\mathrm{Mod}(\mathcal{T}/L)$. This amounts to the same as saying that the pair

$$(\varphi(\vec{x}),\ \varphi\wedge \boldsymbol{f}\vec{x}=\vec{y}) \tag{2}$$

is an ifs; here, $\boldsymbol{f}\vec{x}=\vec{y}$ is an abbreviation of $\bigwedge_{i<n} \boldsymbol{f}_{A_i}(x_i) = y_i$, with $\langle x_i\rangle_{i<n} = \vec{x}$, x_i of sort $(A_i)_{(0)}$, and y_i a (new) variable of sort $(A_i)_{(1)}$. We call an ifs of the form (2) *simple*.

Another method of getting an ifs is to pass from one ifs to an "isomorphic copy" of it. Suppose $(\varphi(\vec{x}), \mu(\vec{x}, \vec{y}))$ is an ifs, and $\gamma(\vec{x}, \vec{x}')$ is an L'-formula that defines, provably in $\mathcal{T}$, a *bijection* from $\varphi(\vec{x})$ to another L'-formula $\varphi'(\vec{x}')$; $\gamma^M:\varphi^M \xrightarrow{\cong} \varphi'^M$ for all

$M \models \mathcal{T}$. Let us define the formula $\mu'(\vec{x}', \vec{y}')$ as the "composite $\mu' = \gamma_{(1)} \circ \mu \circ \gamma^{-1}_{(0)}$",
$\mu'(\vec{x}', \vec{y}') = \exists \vec{x}\vec{y} (\gamma_{(0)}(\vec{x}, \vec{x}') \wedge \mu(\vec{x}, \vec{y}) \wedge \gamma_{(1)}(\vec{y}, \vec{y}'))$, or more generally, take any μ' equivalent to the previous μ' in $\mathcal{T} +_L \mathcal{T}$. This means that the diagram

$$\begin{array}{ccc} \varphi^M & \xrightarrow{\mu^{(M,N,f)}} & \varphi^N \\ {\scriptstyle \gamma^M} \downarrow & & \downarrow {\scriptstyle \gamma^N} \\ \varphi'^M & \xrightarrow[\mu'^{(M,N,f)}]{} & \varphi'^N \end{array}$$

commutes for all $(M, N, f) \models \mathcal{T} +_L \mathcal{T}$. Then, as inspection reveals, (φ', μ') is an ifs again. An ifs (φ', μ') related to (φ, μ) as described is said to be an *isomorphic copy* of (φ, μ).

Our main result says that, in fact, *the two ways indicated are the only ones necessary to get an ifs*, in case certain (mild) assumptions are made on (L, L', T). Namely, we assume that the theory $\mathcal{S} = (L, S)$ *has disjoint sums and quotients of equivalence relations* :

(1) if $\varphi_i(\vec{x}_i)$ $(i<n)$ are L-formulas, then there are L-formulas $\varphi(\vec{x})$, $\gamma_i(\vec{x}_i, \vec{x})$ such that $S \vdash \gamma_i : \varphi_i \longrightarrow \varphi$ (in the expected sense; γ_i defines a function from φ_i to φ), and in fact S proves that the γ_i form the coprojections of a disjoint sum; and

(2) if $\varphi(\vec{x})$, $\rho(\vec{x}, \vec{x}')$ are L-formulas such that S proves that ρ defines an equivalence relation on φ, then there are L-formulas $\psi(\vec{y})$ and $\gamma(\vec{x}, \vec{y})$ such that $S \vdash \gamma : \varphi \longrightarrow \psi$, and S proves that γ is surjective and it maps two elements to the same element iff they are ρ-equivalent (ψ represents the set of the equivalence classes of φ).

1.Theorem. Suppose $\mathcal{T} = (L', T)$ is a theory, $L \subset L'$, and $\mathcal{S} = (L, S)$ with S the set of L-consequences of T. Assume that $\mathcal{S}$ has disjoint sums and quotients of equivalence relations. Then, any $\mathcal{T}/L$-ifs is an isomorphic copy of a simple $\mathcal{T}/L$-ifs.

Condition (1) is an extremely mild one; it holds whenever there is a sort A in L such that S proves $\exists x \in A \exists y \in A (x \neq y)$. The reason why this condition is never noticed is that in model theory one usually deals with complete, one-sorted theories with infinite models, when the condition always holds. Also, any theory may be "inessentially" extended to have disjoint sums.

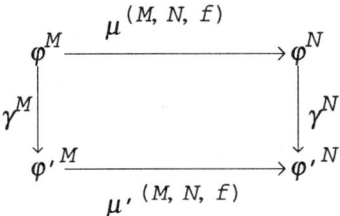

Condition (2) is more substantial, and it is related to Shelah's construction of the many-sorted extension T^{eq} of a theory T; see [Sh], and also [M4]. Any theory of the form T^{eq} satisfies (2), essentially by definition; T^{eq} is also regarded as an "inessential" extension of T. Thus, we see that to obtain a situation when our description of the definable functors on the PC_Δ-category $Mod(\mathcal{T}/L)$ applies, we may take an (almost) arbitrary theory of the form $\mathcal{S}^{eq}$, call its language L, and extend it in a *conservative way* to a theory $\mathcal{T}$ by adding new sorts, relations and axiom.

Let $\nu(z)$ be an L'-formula, with all variables in the sequence z of variables being of sorts in L. We say that ν is *L-invariant* if, for any M, $N \models \mathcal{T}$, and $h: M\restriction L \xrightarrow{\cong} N\restriction L$ we have $M \models \nu[a] \Rightarrow N \models \nu[ha]$ for all tuples a of elements of the corresponding sorts in M.

A way of stating the Beth definability theorem (see [C/K]) is

(BETH) *for every L-invariant* $\nu(z)$ *, there is an L-formula* $\beta(z)$ *such that* $\mathcal{T} \vdash \forall z(\nu(z) \longleftrightarrow \beta(z))$.

The descent theorem 2.2 is *translation-equivalent* via the theory-vs.-pretopos correspondence (described in [M/R]) to the conjunction of 1. and BETH. In particular, with L and $\mathcal{T}$ as above, we have a conservative pretopos morphism $I: S \longrightarrow T$, where T is the Boolean pretopos completion of the theory $\mathcal{T}$, and S is the Boolean pretopos completion of the theory (L, S_0), with S_0 the set of consequences of $\mathcal{T}$ in the language $L_{\omega\omega}$. As it will be explained in detail in Section 11, $Des(I)$, the category of descent data of I (with I considered an arrow in the groupoid-enriched category of Boolean pretoposes), is the same as the category whose objects are the isomorphism functor specifications associated with L and $\mathcal{T}$, and whose arrows are like the isomorphisms of ifs mentioned above, but with the invertibility condition removed. The canonical mapping J of S into $Des(I)$ of the coregular factorization is the mapping that takes an L-formula to the corresponding simple ifs. In the case of a conservative morphism, the quotient-part of the coregular factorization automatically becomes an equivalence. Thus, Theorem 2.2 in the case described says that the map $J: S \longrightarrow Des(I)$ is an equivalence, that is, (i) essentially surjective on objects, and (ii) full and faithful. Here, (i) is expressed directly by Theorem 1. of the present section. It is easily seen that BETH expresses that J is full (surjective) on subobjects, which (for pretoposes) is stronger than being full; faithfulness is automatic by I being conservative.

I will now give a more explicit syntactical form of the theorem, without any assumption on (L, L', T) .

In what follows, x, y, z, z_i, ... denote finite tuples of distinct variables. $\forall x$ denotes simultaneous universal quantification of each variable in x, etc. An *L-variable* is one whose sort is in L. If $z = \langle z_k \rangle_{k<m}$ is a tuple of L-variables, z_k of sort A_k, then $\mathbf{f}z$ denotes $\langle \mathbf{f}_{A_k} z_k \rangle_{k<m}$ (a tuple of terms of the language L^+); recall that the $\mathbf{f}_A$ are the

symbols to denote the components of the L-isomorphism in the theory $\mathcal{T}+_L\mathcal{T}$.

1'. Theorem. Let $(\varphi(x), \mu(x, y))$ be a $\mathcal{T}/L$-ifs, specifying a functor $[\varphi, \mu]:\mathrm{Mod}^*(\mathcal{T}/L) \longrightarrow \mathrm{Set}^*$. Then there are $n<\omega$ and L'-formulas $\gamma_i(z_i, x)$ $(i<n)$, with the z_i all L-variables such that

$$\text{(i)}\ \mathcal{T} \vdash \forall z_i x(\gamma_i(z_i, x) \longrightarrow \varphi(x)) \wedge \forall z \exists^{\leq 1} x \gamma_i(z_i, x)$$

for each $i<n$;

$$\text{(ii)}\ \mathcal{T} \vdash \forall x(\varphi(x) \longrightarrow \bigvee_{i<n} \exists z_i \gamma_i(z_i, x)) \ ;$$

$$\text{(iii)}\ \mathcal{T}+_L\mathcal{T} \vdash \forall xyz_i((\gamma_i)_{(0)}(z_i, x) \longrightarrow (\mu(x, y) \longleftrightarrow (\gamma_i)_{(1)}(\mathbf{f}z_i, y)))$$

for each $i<n$.

The abbreviation $\exists^{\leq 1}x$ stands for "for at most one x". Note that (i) says that each γ_i defines a partial function into the set defined by φ, provably in $\mathcal{T}$. (ii) says that the functions of (i) are jointly surjective onto φ. Finally, (iii) says that in order to find the μ image of any argument x, one may proceed as follows: pick any γ_i-preimage z_i of x, for any $i<n$ for which this is possible; apply the given isomorphism $\mathbf{f}$ (operating on the L-level only) to z_i to get $\mathbf{f}z_i$; finally, take the γ_i image of $\mathbf{f}z_i$.

Almost always, we can, after the fact, make $n=1$, in which case all disjunction signs and subscripts i disappear from the statement of the theorem. This is the case e.g. if $\mathcal{T} \models \exists x \exists y(x \neq_A y)$ for at least one sort A in L (*exercise*); indeed, this is the case when $\mathcal{S}$ has disjoint sums (see above).

It is clear that, in the real parameter $p=(\mathcal{S}, \mathcal{T}, L)$, the statement of the theorem is an arithmetic sentence of the form $\forall n \in \mathbb{N}(\Phi np \longrightarrow \Psi np)$, with both Ψ and Ψ being Σ^0_1-formulas of arithmetic. If p is recursive, we get a (light-face) Π^0_2-sentence in arithmetic; that is, a sentence of the form $\forall n \exists m R(n, m)$, with R a primitive recursive predicate, of elementary arithmetic.

4. BASIC NOTIONS FOR DUALITY

Let us start with what seems to be the most direct way of indicating what is happening in the duality theory we want to develop. We have a theory $\mathcal{T}$ and its groupoid $\mathrm{Mod}^*(\mathcal{T})$ of models. The issue is: can we recover $\mathcal{T}$ from $\mathrm{Mod}^*(\mathcal{T})$ endowed with some "natural" additional structure? Furthermore, we would like to have the additional structure to be *induced* by some structure on the category of sets and isomorphisms, and we would like to do the reconstruction through the maps from $\mathrm{Mod}^*(\mathcal{T})$ to Set that *preserve* the structure.

An easily understood piece of structure that may be expected to work is ultraproducts. We have ultraproducts defined on Set in the form of functors of the shape $\mathrm{Set}^I \to \mathrm{Set}$; and on $\mathrm{Mod}^*(\mathcal{T})$ as functors $\mathrm{Mod}^*(\mathcal{T})^I \to \mathrm{Mod}^*(\mathcal{T})$. Although we have to be careful with the idea of a map $\mathrm{Mod}^*(\mathcal{T}) \to \mathrm{Set}$ preserving ultraproducts (it is not reasonable to require strict preservation; one should have preservation up to specified and natural isomorphisms; see below), it is nevertheless rather clear that the definable functors $\mathrm{Mod}^*(\mathcal{T}) \to \mathrm{Set}$ (see Section 3) do preserve ultraproducts; this is another way of saying that the ultraproduct structure on $\mathrm{Mod}^*(\mathcal{T})$ is Set-induced. What we are aiming at is putting enough Set-induced structure on $\mathrm{Mod}^*(\mathcal{T})$ so that all structure preserving functors $\mathrm{Mod}^*(\mathcal{T}) \to \mathrm{Set}$ turn out to be isomorphic to definable functors (assuming $\mathcal{T}$ has quotients of equivalence relations and disjoint sums; otherwise the enterprise cannot succeed).

The paper [M1] describes how to do this for $\mathrm{Mod}(\mathcal{T})$ instead of $\mathrm{Mod}^*(\mathcal{T})$. From a naive point of view, it is clear that doing this with $\mathrm{Mod}^*(\mathcal{T})$ should be more difficult; we want to recover the same $\mathcal{T}$ from, at least initially, fewer data than before. Nevertheless, the recovery is possible; this is the content of the duality theory of the present paper. The structure to be induced on $\mathrm{Mod}^*(\mathcal{T})$ includes, besides ultraproducts, limit ultrapowers, isomorphism-versions of ultramorphisms of [M1] (certain natural transformations between generalized composites of ultraproduct and limit ultrapower functors), and one more curious piece of structure: certain relations on hom-sets, traces of structure involving non-isomorphism arrows.

Section 1 to 3, the first twenty pages or so, of [M1] are concerned with specifying the *ultrastructure* of Set, the category of sets, with placing $\mathbf{Set}$ into the 2-category of *ultracategories*, and with relating, through a pair of adjoint (2-)functors, the 2-category of pretoposes and that of ultracategories. This whole development is a purely formal concept-formation starting with the two "commuting" structures on Set , the pretopos structure and the ultrastructure; its result is called (in [M1]) the *Stone adjunction* based on those commuting structures. Here "purely formal" means that we only use the formal aspects of those structures; put in another way, the Stone adjunction is a general construction, always available once a pair of "commuting" structures on the same category (or set) is given.

The general notion of Stone adjunction, for set-based structures rather than category-based structures, is described in detail in [J] (but what is called Stone-type duality in [J] is something more special; see "general concrete dualities" in [J] for what we mean here). The more complicated general construction of the Stone adjunction for category-based structures, resulting in a 2-adjunction rather than an ordinary adjunction, does not seem to be written down in the literature. Actually, there is a simple reason why one refrains from giving the general construction. When one or both of the two structures of the "base-"category are given through universal properties, one has a considerable simplification with respect to the general case, the upshot of which is that it is not convenient to use the general framework in these special cases. When both structures are given by universal properties, the situation is simplified to such an extent that it does not require any separate treatment. This is the situation e.g. in the case of the Gabriel-Ulmer duality (see [M/P]), or in the case of the duality for Barr-exact categories (see [M5]). On the other hand, in the ultra-theory [M1], one structure (pretopos) is given by universal properties, but the other (ultrastructure) is not. We find ourselves in an intermediate situation which fits only awkwardly in a general framework.

The complicated element of [M1], the presence of the so-called *ultramorphisms*, is a direct consequence of the circumstance of the ultraproducts not being (directly) defined through universal properties, and it is part of the general framework of Stone adjunction on the 2-categorical level. I emphasize this point in an attempt to convince the reader that the concept of ultramorphism, however involved it may appear, is part of the universal algebra of categories, and therefore, it seems to be inevitable.

The basis for Zawadowski's proof of his descent theorem is the Stone adjunction of pretoposes and ultracategories in [M1]. Similarly, the basis of the proof of the descent theorem for Boolean pretoposes is the Stone adjunction of, on the one hand, Boolean pretoposes, and on the other, *ultragroupoids*. The notion of ultragroupoid is completely specified, in the general framework of Stone adjunctions, by the *standard ultragroupoid* $\mathbf{Set}^*$. As the notation indicates, this is a structure based on the groupoid Set^* of all sets: this is the category whose objects are the (small) sets, and whose arrows are the bijections, with the usual composition.

In a sense, it is fortunate that I did not try to spell out the general concept of 2-dimensional Stone adjunction in [M1] (in fact, I said in [M1] (p. 163) that "we do not have a satisfactory formulation of it"), because the general concept I had in mind at the time of writing

[M1] is not general enough for the present purposes! I will give the Stone adjunction for Boolean pretoposes and ultragroupoids in detail, and I will point out the peculiarities of it that seem to make a suitable general framework difficult to describe.

In order to economize on space, I will try to use as much of [M1] as possible; sometimes formulations like ones in [M2] will be used; the reader is advised to have copies of [M1] and [M2] at hand.

The *ultraproduct functors* on Set are explained on p. 108 of [M1].

With (I, U) an ultrafilter (that is, U an ultrafilter on the set I), A a set, the *ultrapower* A^U is the ultraproduct $\prod_{i\in I} A/U$. $\delta = \delta_A = \delta_A^U : A \longrightarrow A^U$ is the *diagonal mapping* for which $\delta(a) = \langle a\rangle_{i\in I}/U$. For $f:A\longrightarrow B$, $f^U:A^U\longrightarrow B^U$ is the mapping $f^U = \prod_{i\in I} f/U$. By these specifications, we have defined the *ultrapower functor* given by (I, U) :

$$(-)^U : \mathrm{Set} \longrightarrow \mathrm{Set} .$$

If $\vec{U} = \langle U_1, \ldots, U_n\rangle$ is a finite sequence of ultrafilters, we have the iterated ultrapower $A^{\vec{U}} = A^{U_1, \ldots, U_n} = (\ldots((A^{U_1})^{U_2})\ldots)^{U_n}$. We also have the associated "diagonal" map

$$\delta_A^{\vec{U}} = \delta^{U_n}_{A^{U_1}, \ldots U_{n-1}} \circ \ldots \circ \delta^{U_2}_{A^{U_1}} \circ \delta^{U_1}_A : A \longrightarrow A^{\vec{U}} .$$

With $\vec{U} = \langle (I_n, U_n)\rangle_{1\le n<\omega}$ an ω-type sequence of ultrafilters, and with A a set, let $\vec{U}\restriction[1, n] = \langle U_1, \ldots, U_n\rangle$ (empty for $n = 0$), $A_n = A^{\vec{U}\restriction[1, n]}$ for $n < \omega$ (just A for $n = 0$), and $\delta_n = \delta^{U_n}_{A_n} : A_n \longrightarrow A_{n+1}$; we have the ω-type diagram $\langle A_n, \delta_n\rangle_{n<\omega}$; its colimit is the *limit ultrapower* $A^{\vec{U}}$ of A. For specificity, by $A^{\vec{U}}$ we will always mean the canonical representation of the colimit, in which an element of $A^{\vec{U}}$ is an equivalence class of elements of the various ultrapowers involved in the usual way. The notation

$$\delta_A^{\vec{U}, n} : A^{\vec{U}\restriction[1, n]} \longrightarrow A^{\vec{U}}$$

stands for the colimit coprojection; when $n = 0$, n is omitted from the notation.

For $f:A\longrightarrow B$, we have $f^{\vec{U}}:A^{\vec{U}}\longrightarrow B^{\vec{U}}$ defined in the obvious way.

The above specifications add up to the definition of the *limit ultrapower functor* $[\vec{U}]:\mathrm{Set}\longrightarrow\mathrm{Set}$, one for each ultrafilter sequence $\vec{U}$.

By "composing" δ's of the type introduced so far, we get further canonical maps. We single out one type for future use. With $U=(I,U)$ an ultrafilter, and $\langle\vec{U}^i\rangle_{i\in I}$ a family of ultrafilter sequences $\vec{U}^i$, and with the abbreviation $\vec{U}=(\langle\vec{U}^i\rangle_{i\in I},U)$, we put

$$\delta_A^{\vec{U}} \underset{\mathrm{def}}{=} \Big(\prod_{i\in I}\delta_A^{\vec{U}^i}/U\Big)\circ\delta_A^U : A\longrightarrow A^{\vec{U}} \underset{\mathrm{def}}{=} \prod_{i\in I}A^{\vec{U}^i}/U .$$

On Set^*, the groupoid of sets, we impose the ultraproduct functors and the limit ultrapower functors, with their actions restricted to isomorphisms, as additional operations; in addition, we add certain *relations* on arrows in Set^*, a remainder of the full category structure Set; the resulting structure we call the *pre-ultragroupoid of sets*; "pre-" because for the final notion of *ultragroupoid of sets* we need more structure to be introduced later. At the same time, we define the general notion of *pre-ultragroupoid.*

Pre-ultracategories (p.-u.c.'s), *pre-ultrafunctors* (p.-u.f.'s) and *ultratransformations* (u.t.'s) were defined on p. 109 in [M1].

A *pre-ultragroupoid* (p.-u.gr.) $\boldsymbol{S}$ is a p.-u.c. whose underlying category S is a groupoid (all arrows are isomorphisms), together with

(i) a functor (*limit ultrapower* or *ultralimit functor*) $[\vec{U}]=[\vec{U}]_{\boldsymbol{S}}:S\longrightarrow S$ for each ultrafilter sequence $\vec{U}$,

and

(ii) for any ultrafilter $U=(I,U)$, and any family $\langle\vec{U}^i\rangle_{i\in I}$, with each $\vec{U}^i$ an ultrafilter sequence, with abbreviating $(\langle\vec{U}^i\rangle_{i\in I},U)$ as $\vec{U}$, $[U](\langle A^{\vec{U}^i}\rangle_{i\in I})$ as $A^{\vec{U}}$, and similarly for arrows in place of A, a family

$$\vec{E}=\langle E_{A,B}^{\vec{U}}\subset(S(A^{\vec{U}},B))^2 : A,B\in\mathrm{Ob}(S)\rangle$$

of relations $E_{A,B}^{\vec{U}}$, one for each pair of objects A and B, $E_{A,B}^{\vec{U}}$ being a relation on the set of arrows from $A^{\vec{U}}$ to B; $\vec{E}$ is required to satisfy the following invariance condition:

whenever $h:A\xrightarrow{\cong}A'$, $i:B\xrightarrow{\cong}B'$, and in the diagram

$$\begin{array}{ccc} A^{\vec{U}} & \overset{f}{\underset{g}{\rightrightarrows}} & B \\ {\scriptstyle h^{\vec{U}}}\downarrow & & \downarrow{\scriptstyle i} \\ A'^{\vec{U}} & \overset{f'}{\underset{g'}{\rightrightarrows}} & B' \end{array}$$

two squares, one involving f and f', the other involving g and g', commute, then

$(f, g) \in E^{\vec{U}}_{A,\,B}$ iff $(f', g') \in E^{\vec{U}}_{A',\,B'}$.

We need quickly say something to motivate this definition. First of all, there is nothing sacred about the combination $\vec{U} = (\langle \vec{U}^i \rangle_{i \in I}, U)$, although it is the minimum size we actually need. Secondly, $\mathbf{Set}^*$, the *p.-u.gr. of sets*, the "*standard*" p.-u.gr., is the obvious structure as far as the ultraproduct and limit ultrapower functors are concerned, involving restrictions to isomorphisms of those functors defined on the category of sets. It has the following E-relations: for each $\vec{U}$, sets A, B, and isomorphisms $A^{\vec{U}} \overset{f}{\underset{g}{\rightrightarrows}} B$, f and g are in the relation $E^{\vec{U}}_{A,\,B}$ iff the composites $f \circ \delta^{\vec{U}}_A$ and $g \circ \delta^{\vec{U}}_A$ are equal. Clearly, the invariance condition is satisfied.

You see, the map $\delta^{\vec{U}}_A$ is not available in the groupoid Set^*; however, it has an effect on isomorphisms (it may equalize two isomorphisms from its codomain) that we have to recognize in the structure to be adopted.

With $\mathbf{S}$, $\mathbf{S}'$ p.-u.gr.'s, a *pre-ultra*functor* (p.-u*.f.) $X : \mathbf{S} \longrightarrow \mathbf{S}'$ is a p.-u.f. between the underlying p.-u.c.'s, *together with transition isomorphisms* $[X, \vec{U}] : X \circ [\vec{U}]_{\mathbf{S}} \xrightarrow{\cong} [\vec{U}]_{\mathbf{S}'} \circ X$, one associated with each ultrafilter sequence $\vec{U}$, and *such that* X preserves the E-relations in the natural sense. To see what this means, note that X induces, also using its transition isomorphisms, a mapping

$$X^{\#} : S(A^{\vec{U}}, B) \longrightarrow S'((XA)^{\vec{U}}, XB)$$

for each appropriate value of the parameters involved. To say that X preserves the E-relations is to say that

$$(f, g) \in (E^{\vec{U}}_{A,\,B})_{\mathbf{S}} \Longrightarrow (X^{\#}f, X^{\#}g) \in (E^{\vec{U}}_{XA,\,XB})_{\mathbf{S}'} \ . \qquad (1)$$

The effect of the mapping $X^{\#}$ is described by saying that

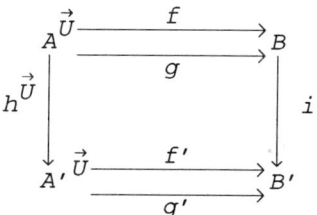

$$\begin{array}{ccc} (XA)^{\vec{U}} & \xrightarrow{\;X^{\#}f\;} & XB \\ {\scriptstyle [X,\,\vec{U}]_A}\Big\uparrow & & \Big\uparrow{\scriptstyle 1_B} \\ X(A^{\vec{U}}) & \xrightarrow[\;Xf\;]{} & XB \end{array}$$

commutes; $[X,\vec{U}]_A \underset{\mathrm{def}}{=} [U](\langle [X,\vec{U}^i]\rangle_{i\in I}) \circ [X,U]_{\langle A^{\vec{U}^i}\rangle_{i\in I}}$.

Given a family $\mathcal{X}$ of p.-u$\overset{*}{.}$f.'s $X:\boldsymbol{S}\longrightarrow\boldsymbol{S}'_X$ with the same domain $\boldsymbol{S}$, we say that $\mathcal{X}$ is *relation-conservative* if the implications (1) can be jointly reversed, that is, if for any specific pair (f,g) the right-hand sides are true for all $X\in\mathcal{X}$, then the left side is true as well.

An *ultra** *transformation* (u$\overset{*}{.}$t.) $\sigma:X\longrightarrow Y$ between p.-u$\overset{*}{.}$f.'s X and Y, $\boldsymbol{S}\underset{Y}{\overset{X}{\rightrightarrows}}\boldsymbol{S}'$, is a u.t. between the underlying p.-u.f.'s satisfying the additional condition that the diagram

$$\begin{array}{ccc} X\circ[\vec{U}]_{\boldsymbol{S}} & \xrightarrow{\;\sigma\circ[\vec{U}]_{\boldsymbol{S}}\;} & Y\circ[\vec{U}]_{\boldsymbol{S}} \\ {\scriptstyle [X,\,\vec{U}]}\Big\downarrow & & \Big\downarrow{\scriptstyle [Y,\,\vec{U}]} \\ [\vec{U}]_{\boldsymbol{S}'}\circ X & \xrightarrow[\;[\vec{U}]_{\boldsymbol{S}'}\circ\sigma\;]{} & [\vec{U}]_{\boldsymbol{S}'}\circ Y \end{array} \tag{2}$$

commutes for each ultrafilter sequence $\vec{U}$.

We write $\mathrm{Hom}(\boldsymbol{S},\boldsymbol{S}')$ for the category (groupoid) of all p.-u$\overset{*}{.}$f.'s from $\boldsymbol{S}$ to $\boldsymbol{S}'$, with the u$\overset{*}{.}$t.'s as morphisms (composition being that of natural transformations).

Since every morphism in a p.-u.gr. is an isomorphism, the components of an u$\overset{*}{.}$t. are isomorphisms; in particular (following the above notation with $\boldsymbol{S}'=\mathbf{Set}^*$, the p.-u.gr. of sets) u$\overset{*}{.}$t.'s between p.-u$\overset{*}{.}$f.'s into $\mathbf{Set}^*$ have all their components isomorphisms. However, in this case there is a natural notion of "ultratransformation" with components arbitrary arrows in Set , the reason being that the ultraproduct and limit ultrapower functors are defined on the whole of Set , rather than just Set^* .

Let $\boldsymbol{S}$ be an arbitrary p.-u.gr., let $\boldsymbol{S}\underset{Y}{\overset{X}{\rightrightarrows}}\mathbf{Set}^*$ be a pair of p.-u$\overset{*}{.}$f.'s into the distinguished p.-u.gr. of sets. A *lax ultra** *transformation* (lax u$\overset{*}{.}$t.) $\sigma:X\longrightarrow Y$ is a natural transformation between the underlying functors $S \underset{Y}{\overset{X}{\rightrightarrows}} \mathtt{Set}$ (note the dropping of the $*$!) satisfying the appropriate commutativity conditions concerning the transition isomorphisms attached to the ultraproduct and limit ultrapower functors. That is, σ is a u.t. between the underlying p.-u.f.'s between the p.-u.c.'s $\mathbf{S}$ and **Set**, also satisfying that the diagram (2), with $[\vec{U}]_{\mathbf{S}'}$ in the lower line replaced by the ultrapower functor $[\vec{U}] : \mathtt{Set} \longrightarrow \mathtt{Set}$, commutes, for all appropriate values of the parameters involved. The components of a lax $\overset{*}{\mathrm{u.t.}}$ are not necessarily bijections.

We will write $\mathrm{Hom}(\mathbf{S}, \mathbf{Set})$ for the category of p.-$\overset{*}{\mathrm{u}}$.f.'s from $\mathbf{S}$ to $\mathbf{Set}^*$, with the lax $\overset{*}{\mathrm{u}}$.t.'s as morphisms. Note that $\mathrm{Hom}(\mathbf{S}, \mathbf{Set}^*)$ is something different; the latter is a subcategory of the former, with the morphisms in the latter being all isomorphisms .

A p.-$\overset{*}{\mathrm{u}}$.f. is *strict* if all transition isomorphisms of it are identities. The forgetful functor

$$\mathrm{Hom}(\mathbf{S}, \mathbf{S}') \longrightarrow \mathrm{Hom}(S, S')$$

is called a *quasi inclusion* in [M1] as well as here; note that, with $\mathbf{S}$ a p.-u.gr. and $\mathbf{S}'$ either a p.-u.gr. or $\mathbf{S}' = \mathbf{Set}$, this makes sense; in the first case, we have groupoids, in the second not.

The definition of composition of p.-$\overset{*}{\mathrm{u}}$.f.'s is the obvious extension of that for p.-u.f.'s given on p.110 [M1]. Just as in *loc.cit.*, we then have the functor

$$\mathrm{Hom}(X, \mathbf{K}'') = (\ )\circ X : \mathrm{Hom}(\mathbf{K}', \mathbf{K}'') \longrightarrow \mathrm{Hom}(\mathbf{K}, \mathbf{K}'')$$

whenever $\mathbf{K}, \mathbf{K}'$ are p.-u.gr.'s , $X:\mathbf{K} \longrightarrow \mathbf{K}'$, and either $\mathbf{K}''$ is a p.-u.gr., or $\mathbf{K}'' = \mathbf{Set}$.

Next, we introduce the modification of the notion of ultramorphism we need. Some terminology from [M1] and [M2] will be repeated.

Here and later, we will frequently deal with formal entities that are tuples such as (γ, I, U, g) , and we need a notation picking out the components of the tuple in dependence on the tuple. If u denotes the tuple above, then γ_u , I_u , U_u and g_u will mean the components in this order. The notation will be used also in one level deeper; e.g., if $\ell = (\gamma, \gamma', \langle U_n \rangle_{1 \le n < \omega})$, then $U_{n\ell}$ stands for U_n . In this notation, the character in the main line is invariable; it is the name of a function; the subscript(s) is (are) argument(s) of the function. There will be one exception to this invariability. The first component of our tuples will (usually) be nodes in a graph. If the name of that graph is Γ , we will use γ_u as in the example; but if the name of the graph is Λ or Σ , we switch to λ_u , σ_u , respectively. We hope this will not lead to confusion.

Let Γ be a graph. An *ultraproduct specification*, or more simply, an *ultraproduct*, in Γ is a quadruple $u = (\gamma, I, U, g)$ where $\gamma \in |\Gamma|$, (I, U) is an ultrafilter, and $g : I \longrightarrow |\Gamma|$.

An *ultralimit* (or limit ultrapower) (*specification*) in Γ is an entity of the form

$\ell = (\gamma, \gamma', \vec{U})$ with $\gamma, \gamma' \in |\Gamma|$, and with $\vec{U}$ an ω-type sequence of ultrafilters.

An *ultra*graph* (u*g.) Γ is a graph, also denoted by Γ, together with a set $\mathcal{U}_\Gamma$ of ultraproduct specifications in Γ, and a set $\mathcal{L}_\Gamma$ of ultralimit specifications in Γ. When Γ is an ultra*graph, an ultraproduct *in* Γ means one in $\mathcal{U}_\Gamma$.

An ultra*graph will always be a small one; this is to be emphasized since it would be natural to consider the underlying ultra*graph of a possibly large p*-u.c.; we will refrain from doing so.

Let Γ be an u*g., $\mathcal{S}$ a p.-u*c. An *ultra*diagram* (u*d.) $\mathcal{A}:\Gamma \longrightarrow \mathcal{S}$ is a diagram $\mathcal{A}:\Gamma \longrightarrow S$ between the underlying graphs, together with *transition isomorphisms*

$$[\mathcal{A}, u] : \mathcal{A}(\gamma_u) \xrightarrow{\cong} [U_u]\langle \mathcal{A}g_u i\rangle_{i \in I_u},$$

one for each $u \in \mathcal{U}_\Gamma$, and

$$[\mathcal{A}, \ell] : \mathcal{A}(\gamma_\ell) \xrightarrow{\cong} [\vec{U}_\ell](\mathcal{A}((\gamma')_\ell)),$$

one for each $\ell \in \mathcal{L}_\Gamma$.

A *morphism* $\varphi:\mathcal{A} \longrightarrow \mathcal{B}$ of u*d.'s $\Gamma \underset{\mathcal{B}}{\overset{\mathcal{A}}{\rightrightarrows}} \mathcal{S}$ is a natural transformation of diagrams into a category in the usual sense, satisfying the natural compatibility conditions relative to the transition isomorphisms in $\mathcal{A}$ and $\mathcal{B}$; see [M2], bottom of page 181. $\mathrm{Hom}(\Gamma, \mathcal{S})$ is the category (groupoid) of all u*d.'s from Γ to the p-u.gr. $\mathcal{S}$ and their morphisms.

If in the last definition, $\mathcal{S}$ is $\mathbf{Set}^*$, we have the notion of *lax morphism* $\varphi:\mathcal{A} \longrightarrow \mathcal{B}$ whose components are arbitrary arrows in Set, rather than just isomorphisms (all the components of an ordinary morphism of u*d.'s are automatically isomorphisms). $\mathrm{Hom}(\Gamma, \mathbf{Set})$ is the category of all u*d's from Γ to $\mathbf{Set}^*$, and their lax morphisms.

Let $\mathcal{S}$ be a p.-u.gr., Γ an u*g., γ_1 and γ_2 two nodes in Γ; ev_γ is the functor $\mathrm{ev}_\gamma : \mathrm{Hom}(\Gamma, \mathcal{S}) \longrightarrow S$ defined as evaluation at γ. An *ultra*morphism* (u*m.) *of type* $(\Gamma, \gamma_1, \gamma_2)$ *in* $\mathcal{S}$ is a natural transformation $\mathrm{ev}_{\gamma_1} \longrightarrow \mathrm{ev}_{\gamma_2}$. Note that all components of an u*m. are isomorphisms.

Let us now consider $\mathbf{Set}^*$ instead of $\mathcal{S}$. A *full ultra*morphism of type* (Γ, k, ℓ) *in* $\mathbf{Set}^*$ is a natural transformation $\delta : \mathrm{ev}_k \longrightarrow \mathrm{ev}_\ell$, with $\mathrm{ev}_k : \mathrm{Hom}(\Gamma, \mathbf{Set}) \longrightarrow \mathrm{Set}$ the evaluation, whose components are not just isomorphisms!, and similarly for ev_ℓ, *such that* all components of δ are isomorphisms (note that the difference with respect to the non-full version is a more stringent naturality condition).

We also occasionally use the old notion of ultramorphism ([M1] and [M2]), with the modification that limit ultrapowers are also involved (and colimits are not, contrary to [M2]); the meanings should be clear.

Let $X: \mathcal{S} \longrightarrow \mathcal{S}'$ be a p.-u*.f. between p.-u.gr.'s, let v, v' be u*.m.'s of the same type, on $\mathcal{S}$, $\mathcal{S}'$, resp. We have the natural notion of X carrying v into v'; the definition is a direct copy of the corresponding notion for ultramorphisms on page 182 of [M2].

An u*.d. $\mathcal{A}: \Gamma \longrightarrow \mathcal{S}$ is *strict* if all its transition isomorphisms are identities. When dealing with u*.d.'s in detail, it is a welcome simplification if one can restrict attention to strict ones. This is indeed possible provided the u*.g.'s that serve as domains are of a special kind. Let Γ be an u*.g. Consider the following relation $\prec$ on the nodes of Γ. $\gamma' \prec \gamma$ iff there is an ultraproduct or limit ultrapower specification with γ its principal node (the first component of the specification), and with γ' one of its subsidiary nodes: in case of an ultraproduct u, $\gamma' = g_u(i)$ for some $i \in I_u$; in case of an ultralimit ℓ, $\gamma' = \gamma'_\ell$. We call Γ *stratified* if no two distinct specifications have the same principal node, and the relation $\prec$ is well-founded (no $\prec$-descending infinite sequence).

If Γ is stratified, then every $\mathcal{A}: \Gamma \longrightarrow \mathcal{S}$ is isomorphic (in $\mathrm{Hom}(\Gamma, \mathcal{S})$) to a strict one; in fact, there is a unique strict $\mathcal{B}: \Gamma \longrightarrow \mathcal{S}$ such that there is an isomorphism $\varphi: \mathcal{A} \longrightarrow \mathcal{B}$ which is the identity on the *free* nodes, that is, which are not the principal node of any specification; in fact, φ is unique as well. Thus, the full subcategory $\mathrm{Hom}_s(\Gamma, \mathcal{S})$ of $\mathrm{Hom}(\Gamma, \mathcal{S})$ consisting of the strict u*.d.'s is actually equivalent to $\mathrm{Hom}(\Gamma, \mathcal{S})$ via the inclusion.

It is possible to get rid of all but the strict ultra*functors too, but this requires some very artificial stratification assumptions on the effect of ultraproducts in our categories.

Note that non-strict u*.d.'s come in as composites of a (strict) u*.d. and a (non-strict) u*.f.

In [M1] only the strict versions of ultragraph and ultradiagram are used, and the notion of ultragraph is further restricted. This circumstance requires some special moves in defining the notion of a p.-u.f. carrying one ultramorphism into another. In this paper, just as in [M2], we stick to the general notion.

Let Δ be an arbitrary class of full ultra*morphisms on $\mathbf{Set}^*$. A Δ-*ultragroupoid* (Δ-u.gr.) is a p.-u.gr. $\mathcal{S}$, with an u*.m $v_{\mathcal{S}}$ on $\mathcal{S}$ of the same type as v, for each $v \in \Delta$. A Δ-*ultra*functor* $X: \mathcal{S} \longrightarrow \mathcal{S}'$ between Δ-u.gr.'s is a p.-u*.f. that carries $v_{\mathcal{S}}$ into $v_{\mathcal{S}'}$, for each $v \in \Delta$. For Δ-u.gr.'s $\mathcal{S}$, $\mathcal{S}'$, $\mathrm{Hom}(\mathcal{S}, \mathcal{S}')$ denotes the category whose objects are the Δ-u*.f.'s, and whose arrows are the u*.t.'s (unchanged from the previous definition).

$\mathrm{Hom}(\mathcal{S}, \mathbf{Set})$ has as objects the Δ-u*.f.'s, and arrows the full u*.t.'s. $\mathbf{Set}^*$, somewhat ambiguously, will denote the *Δ-ultragroupoid of sets and functions*, in which, of course, $\nu_{\mathbf{Set}^*} = \nu$ ($\nu \in \Delta$). In fact, in the next section, when dealing with these notions in a general way, we will suppress all mention of Δ, although a choice of Δ will be presupposed; we will talk about u.gr.'s instead of Δ-u.gr.'s, etc.

5. THE STONE–TYPE ADJUNCTION FOR BOOLEAN PRETOPOSES AND ULTRAGROUPOIDS

We give a series of definitions paralleling (i)(a)-(vi)(b) in Section 2 and their Δ-subscripted versions in Section 3, of[M1]. Let us fix an arbitrary class Δ of full $\mathrm{u}^*\!.\mathrm{m}$.'s in $\mathbf{Set}^*$ as at the end of the last section. The definitions will introduce notions depending on Δ; however, the dependence on Δ will not be shown in the notation. Later, in sections 6 and 7, we will specify a definite class Δ in which we are interested in.

To abbreviate formulas, S now denotes Set (written SET in [M1]), $\mathcal{S}$ is Set as a *Boolean* pretopos, $\mathcal{T} = T$, $\mathcal{T}' = T'$ arbitrary Boolean pretoposes, $\boldsymbol{K}$, $\boldsymbol{K'}$ u.gr.'s with underlying groupoids K, K', resp. $\mathbf{S}$, abbreviating $\mathbf{Set}$, will be used in the context $\mathrm{Hom}(-,\mathbf{S})$, with the blank filled by a u.gr., or a morphism of such, with the meanings explained above. $\mathbf{S}^*$ means the u.gr. of sets (see above). We will use the superscript * on Hom to mean the groupoid part of the category originally intended. Thus, with T a category, $\mathrm{Hom}^*(T, S)$ means the groupoid of all set-valued functors with domain T.

The twelve statements $(\mathrm{i})^*(\mathrm{a})$-$(\mathrm{vi})^*(\mathrm{b})$ in question, each defining a new piece of structure, are almost the same as the corresponding ones in [M1]. One has to read "Boolean pretopos" for "pretopos", "u.gr." for "p.-u.c.", "$\mathrm{u}^*\!.\mathrm{f}$." for "p.-u.f." . One has to put the superscript * on some Hom's (including some boldface **Hom**'s) and S's.

(i)*(a) $\mathrm{Hom}(K, \mathrm{S})$ is a Boolean pretopos, denoted $\mathcal{H}om(K, \mathrm{S})$, such that we have a factorization

$$\begin{array}{ccc} K & \xrightarrow{\quad \mathrm{ev}_K \quad} & \mathrm{Hom}(\mathrm{Hom}(K, \mathrm{S}), \mathrm{S}) \\ & \searrow \qquad\qquad \nearrow_{\text{inclusion}} & \\ & \mathrm{Hom}(\mathcal{H}om(K, \mathrm{S}), \mathcal{S}) & \end{array}$$

Here, K is an arbitrary groupoid. The assertion says that S^K is a Boolean pretopos, with "pointwise" operations.

(i)*(b) $\mathrm{Hom}^*(T, \mathrm{S})$ is made into a u.gr., denoted $\mathbf{Hom}^*(T, \mathrm{S})$, in a unique way such that we have a factorization

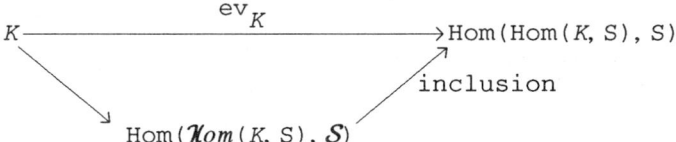

$$\begin{array}{ccc} T & \xrightarrow{\mathrm{ev}_T} & \mathrm{Hom}(\mathrm{Hom}^*(T,\mathrm{S}),\mathrm{S}) \\ \downarrow & & \uparrow \text{quasi-inclusion} \\ \mathrm{Hom}_{\mathbf{S}}(\mathbf{Hom}^*(T,\mathrm{S}),\mathbf{S}) & \xrightarrow[\text{inclusion}]{} & \mathrm{Hom}(\mathbf{Hom}^*(T,\mathrm{S}),\mathbf{S}) \end{array}$$

and such that the evaluation functors $\mathrm{ev}_A:\mathrm{Hom}^*(T,\mathrm{S})\longrightarrow\mathrm{S}$ $(A\in T)$ (which, by the part of the description given so far, are all strict u.f.'s $\mathrm{ev}_A:\mathbf{Hom}^*(T,\mathrm{S})\longrightarrow\mathbf{S}^*$), form a jointly relation-conservative family. [The *fullness* of the u.m.'s is needed for the construction!]

(ii)*(a) With $q:\mathrm{Hom}(\boldsymbol{K},\mathbf{S})\longrightarrow\mathrm{Hom}(K,\mathrm{S})$ the quasi-inclusion, define

$$\mathrm{ev}'_K: K\longrightarrow\mathrm{Hom}^*(\mathrm{Hom}(\boldsymbol{K},\mathbf{S}),\mathrm{S})$$

as the composite

$$K\xrightarrow{\mathrm{ev}_K}\mathrm{Hom}^*(\mathrm{Hom}(K,\mathrm{S}),\mathrm{S})\xrightarrow{\mathrm{Hom}(q,\mathrm{S})}\mathrm{Hom}^*(\mathrm{Hom}(\boldsymbol{K},\mathbf{S}),\mathrm{S}) .$$

Then ev'_K is made into a u.f.

$$\mathrm{ev}_{\boldsymbol{K}}:\boldsymbol{K}\longrightarrow\mathbf{Hom}^*(\mathrm{Hom}(\boldsymbol{K},\mathbf{S}),\mathrm{S})$$

by the transition isomorphisms $[\mathrm{ev}_{\boldsymbol{K}}, U]$ defined by

$$([\mathrm{ev}_K, U]_{\langle M_i\rangle_{i\in I}})_X = [X, U]_{\langle M_i\rangle_{i\in I}}$$

((I, U) an ultrafilter, $M_i\in K$ for $i\in I$, $X\in\mathrm{Hom}(\boldsymbol{K},\mathbf{S})$),

and similarly for the transition isomorphisms for limit ultrapowers.

(ii)*(b) With $i:\mathrm{Hom}^*(\mathcal{T},\mathcal{S})\longrightarrow\mathrm{Hom}^*(T,\mathrm{S})$ the inclusion, define

$$\mathrm{ev}'_T: T\longrightarrow\mathrm{Hom}(\mathrm{Hom}^*(\mathcal{T},\mathcal{S}),\mathrm{S})$$

as the composite

$$T\xrightarrow{\mathrm{ev}_T}\mathrm{Hom}(\mathrm{Hom}^*(T,\mathrm{S}),\mathrm{S})\xrightarrow{\mathrm{Hom}(i,\mathrm{S})}\mathrm{Hom}(\mathrm{Hom}^*(\mathcal{T},\mathcal{S}),\mathrm{S}) .$$

Then ev'_T is a pretopos functor, denoted

$$\mathrm{ev}_{\mathcal{T}}:\mathcal{T}\longrightarrow\mathcal{H}om(\mathrm{Hom}^*(\mathcal{T},\mathcal{S}),\mathrm{S}) .$$

(iii)*(a) (functoriality of $\mathrm{Hom}(-,S)$) For any $X:K\longrightarrow K'$, the functor $\mathrm{Hom}(X,S) = (-)\circ X : \mathrm{Hom}(K',S)\longrightarrow \mathrm{Hom}(K,S)$ is a pretopos functor

$$\mathcal{H}om(X,S) : \mathcal{H}om(K',S)\longrightarrow \mathcal{H}om(K,S) .$$

(iii)*(b) (functoriality of $\mathbf{Hom}^*(-,S)$) For any $M:T\longrightarrow T'$, the functor $\mathrm{Hom}^*(M,S) = (-)\circ M : \mathrm{Hom}^*(T',S)\longrightarrow \mathrm{Hom}^*(T,S)$ is a strict u*f.

$$\mathbf{Hom}^*(M,S) : \mathbf{Hom}^*(T',S)\longrightarrow \mathbf{Hom}^*(T,S) .$$

(iv)*(a) $\mathrm{Hom}(\boldsymbol{K},\mathbf{S})$ is a Boolean pretopos, denoted $\mathcal{H}om(\boldsymbol{K},\mathbf{S})$ such that the quasi-inclusion $\mathcal{H}om(\boldsymbol{K},\mathbf{S})\longrightarrow \mathcal{H}om(K,S)$ is a conservative pretopos functor.

(iv)*(b) $\mathrm{Hom}^*(\mathcal{T},\mathcal{S})$ is made into an u.gr., denoted $\mathbf{Hom}^*(\mathcal{T},\mathcal{S})$, such that the inclusion $\mathrm{Hom}^*(\mathcal{T},\mathcal{S})\longrightarrow \mathrm{Hom}^*(T,S)$ is a strict relation-conservative u*f.

$$\mathrm{inclusion} : \mathbf{Hom}^*(\mathcal{T},\mathcal{S})\longrightarrow \mathbf{Hom}^*(T,S) .$$

(v)*(a) (functoriality of $\mathcal{H}om(-,S)$) For any u*f. $X:\boldsymbol{K}\longrightarrow \boldsymbol{K}'$, the induced functor $\mathrm{Hom}(X,\mathbf{S}):\mathrm{Hom}(\boldsymbol{K}',\mathbf{S})\longrightarrow \mathrm{Hom}(\boldsymbol{K},\mathbf{S})$ is a pretopos functor

$$\mathcal{H}om(X,\mathbf{S}) : \mathcal{H}om(\boldsymbol{K}',\mathbf{S})\longrightarrow \mathcal{H}om(\boldsymbol{K},\mathbf{S}) .$$

(v)*(b) (functoriality of $\mathbf{Hom}^*(-,\mathcal{S})$) For any pretopos functor $M:\mathcal{T}\longrightarrow \mathcal{T}'$, the induced functor $\mathrm{Hom}^*(M,\mathcal{S}):\mathrm{Hom}^*(\mathcal{T}',\mathcal{S})\longrightarrow \mathrm{Hom}^*(\mathcal{T},\mathcal{S})$ is a strict u*f.

$$\mathbf{Hom}^*(M,\mathcal{S}) : \mathbf{Hom}^*(\mathcal{T}',\mathcal{S})\longrightarrow \mathbf{Hom}^*(\mathcal{T},\mathcal{S}) .$$

(vi)*(a) We construct the following diagram:

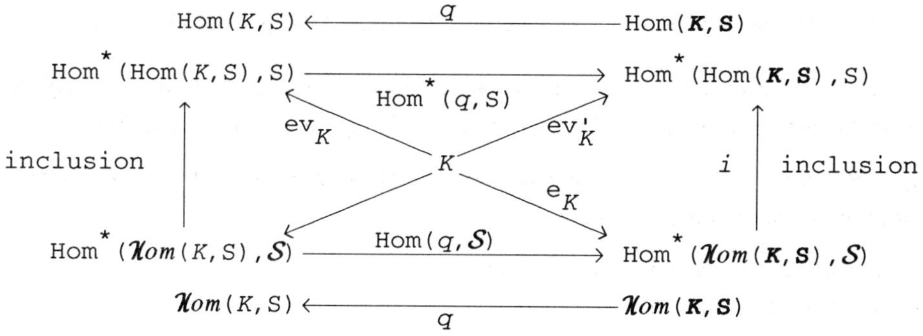

The left and the top triangles are given by (i)*(a) and (ii)*(a), resp., while e_K is being defined by the bottom triangle. Since the square commutes for general reasons, the right triangle commutes too. By (iv)*(b), i is a strict relation-conservative $u^*f.$

$$i : \mathbf{Hom}^*(\mathcal{T}, \mathcal{S}) \longrightarrow \mathbf{Hom}^*(T, S)$$

for $\mathcal{T} = \mathrm{Hom}(\boldsymbol{K}, \mathbf{S})$; also, by (ii)*(a), we have the $u^*f.$

$$\mathrm{ev}_{\boldsymbol{K}} : \boldsymbol{K} \longrightarrow \mathbf{Hom}^*(T, S) ,$$

whose functor part is ev'_K. It follows that e_K is made a $u^*f.$, denoted

$$\eta_{\boldsymbol{K}} : \boldsymbol{K} \longrightarrow \mathbf{Hom}^*(\mathcal{H}om(\boldsymbol{K}, \mathbf{S}), \mathcal{S})$$

in a unique way so that $i \circ \eta_{\boldsymbol{K}} = \mathrm{ev}_{\boldsymbol{K}}$.

(vi)*(b) By the dual procedure, we define the pretopos functor

$$\varepsilon_{\mathcal{T}} : \mathcal{T} \longrightarrow \mathcal{H}om(\mathbf{Hom}^*(\mathcal{T}, \mathcal{S}), \mathbf{S})$$

$\varepsilon_{\mathcal{T}}$ is an "evaluation functor":

$$\begin{array}{lll} \varepsilon_{\mathcal{T}}(A)(M) = M(A) & & (A \in T,\ M \in \mathrm{Hom}(\mathcal{T}, \mathcal{S})) \\ \varepsilon_{\mathcal{T}}(A)(h) = h_A & & (h \in \mathrm{Arr}(\mathrm{Hom}(\mathcal{T}, \mathcal{S}))) \\ \varepsilon_{\mathcal{T}}(f)_M = M(f) & & (f \in \mathrm{Arr}(T)) . \end{array}$$

$\mathcal{BP}^* = \mathcal{B}oole\mathcal{P}retop^{\mathrm{iso}}_{\mathcal{U}_1}$ is the groupoid-enriched category of all Boolean pretoposes (in the universe $\mathcal{U}_1$): its objects are the Boolean pretoposes, its arrows the pretopos functors, 2-arrows all natural isomorphisms. The composition structure is the ordinary one.

$\mathbf{UG} = \mathbf{UltraGroupoid}_{\mathcal{U}_1}$ is the groupoid-enriched category of all u.gr.'s : its objects are the u.gr.'s, its arrows the $u^*f.$'s, 2-arrows all $u^*t.$'s (whose components are isomorphisms). The composition structure is a straightforward one, although it does require definition. Composition of $u^*f.$'s is defined as on page 110 of [M1]. Composition of 2-arrows is that of composition of natural transformations.

In particular, both $\mathcal{BP}^*$ and $\mathbf{UG}$ are 2-categories.

We have the 2-functors

$$\mathcal{BP}^{*\mathrm{op}} \underset{F \,=\, \mathcal{H}om(-,\,\mathbf{Set})}{\overset{G \,=\, \mathbf{Hom}^*(-,\,\mathcal{S}et)}{\rightleftarrows}} \mathbf{UG}\,.$$

The effects of G on objects and arrows are defined in (i)*(b) and (iii)*(b), resp. For

$$\begin{array}{ccc} & \mathcal{T} & \\ P\big\downarrow & \xrightarrow[\cong]{h} & \big\downarrow Q \\ & \mathcal{T}' & \end{array}$$

Gh is defined so that $(Gh)_M = M \circ h$ for every $M: \mathcal{T}' \longrightarrow \mathcal{S}et$. The effects of F on objects and arrows are defined in (i)*(a) and (iii)*(a), resp. The effect on 2-arrows is defined similarly to G.

The functors η_K, ε_T form the components of 2-natural transformations

$$\eta : 1_{\mathbf{UG}} \longrightarrow G \circ F,$$

$$\varepsilon : F \circ G \longrightarrow 1_{\mathcal{BP}^{*\mathrm{op}}}.$$

In fact, $(F, G, \eta, \varepsilon)$ is a 2-adjunction (F is left adjoint to G), that is, we have the "triangle" identities

$$(G\varepsilon) \circ \eta_G = 1_G,$$

$$(F\eta) \circ \varepsilon_F = 1_F.$$

Let us exhibit the form of the *transpose* of an arrow. As usual, if $\mathcal{T} \in \mathcal{BP}$, $\boldsymbol{K} \in \mathbf{UG}$ and $X: \boldsymbol{K} \longrightarrow G(T)$ in $\mathbf{UG}$, then the *transpose* of X is the composite $X^{\#} = F(X) \circ \varepsilon_{\mathcal{T}}: T \longrightarrow F(\boldsymbol{K})$. We have the formulas

$$\begin{array}{lll} X^{\#}(A)(M) = (XM)(A) & & (A \in T,\ M \in K) \\ X^{\#}(A)(h) = (Xh)_A & & (h \in \mathrm{Arr}(K)) \\ X^{\#}(f)_M = (XM)(f) & & (f \in \mathrm{Arr}(T))\,. \end{array}$$

If X is an inclusion, these are the same as the formulas for $\varepsilon_{\mathcal{T}}$ in (vi)*(b).

We have not said much about the proofs of the various statements in this section; they are all routine, "purely formal" verifications, depending only on general ("universal algebraic") aspects of the notions involved. Moreover, the statements are closely parallel to statements in

[M1]; see also Section 8 of [M1]. The resulting 2-adjunction is what we want to call the "Stone-type adjunction" for Boolean pretoposes and ultragroupoids; note the hidden parameter Δ in all the above, which, of course, has to be specified to have something definite.

The interest of the Stone-adjunction is that it has "duality-potentials"; only the duality theory will reveal why e.g. we bothered about the $\mathcal{E}$-relations; the adjunction would be valid just as well without them!

Despite all the lack of novelty in this section, there is something strange about the Stone-adjunction it has given us. What is the naive idea about the Stone adjunction? We have a category $\mathcal{S}$, say; namely, $\mathcal{S} = \mathtt{Set}$ in several interesting cases; then we have two sets of operations, or as here, even relations (that is not the strange thing!) on $\mathcal{S}$; and we have the basic fact of every operation (relation) in one set commuting with any in the other set. But what do we have in our present case? First of all, it is only the groupoid of sets, $\mathcal{S}^{\star}$, that is common to both sides; so we should be talking about $\mathcal{S}^{\star}$ as the schizophrenic object. Now, on the "space" side we have the ultraproduct functors meant as acting on the bijections only. But when we say that the ultraproducts commute with the pretopos structure on the "algebra" side, we in fact mean to take into account the effect of ultraproducts on non-isomorphisms (e.g., product projections) as well; this is the only way to make a good sense of the "commutation". *Where* is the effect of ultraproducts on non-bijections: on the algebra side or on the space side? Obviously, on neither. The "commutation" is not a *condition*; it is an additional *structure*.

Following the naive idea of "Stone-adjunction", it would be easy, albeit lengthy, to explain the general idea precisely; it would indeed cover all the other cases considered so far, e.g. the one in [M1]; however, it would not cover the new case in this paper. I did not see how to give a reasonable, general enough definition of a 2-dimensional Stone adjunction until Robert Paré pointed out to me that one could obtain one using *double categories*.

6. THE SYNTAX OF SPECIAL ULTRAMORPHISMS

This and the next sections are devoted to the construction of a certain class of ultra*morphisms, the *special ultra*morphisms*. They are derived from the "special ultramorphisms" introduced (although not so named) in [M2] (the special ultramorphisms are implicit in [M1] already). To form an idea about these things, recall the classical construction that starts with an elementary embedding $h_0 : M_0 \to M_1$, and constructs the (infinite) commutative diagram

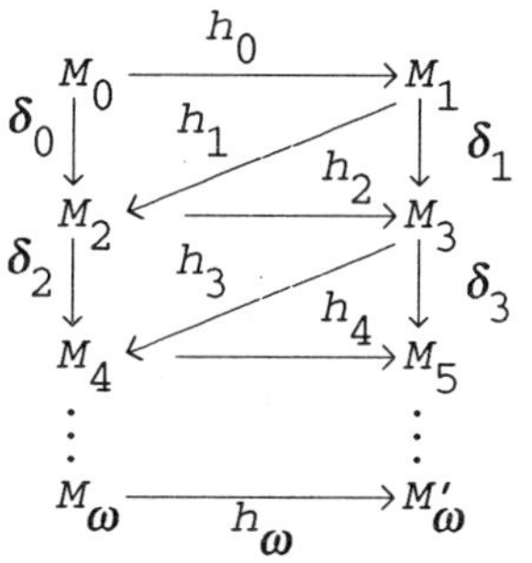

where each M_{k+2} is an ultrapower of M_k, $\delta_k : M_k \to M_{k+2}$ is the corresponding canonical embedding, and h_ω is the resulting isomorphism between the colimits (unions of chains) M_ω, M'_ω of the even, respectively, odd numbered M_k (so that $\delta_{k+1,\,\omega} \circ h_k = h_\omega \circ \delta_{k,\,\omega}$ for k even, with $\delta_{n,\,\omega}$ the colimit injection (inclusion) of M_n into the corresponding colimit). h_ω is an isomorphism "lifting" h_0.

Note that the instances of ultramorphisms on $\text{Mod}(T)$ are elementary embeddings. In fact, when one applies the special ultramorphism construction of [M1] and [M2] to diagrams all whose arrows are isomorphisms, one still gets, in general, non-isomorphism embeddings of models. We will associate an ultra*morphism with each special ultramorphism; it will have instances that are isomorphism liftings in the above sense of the instances of the original ultramorphism. Note that there is a great deal of freedom in choosing the maps h_k in the above construction. In our case, they have to be chosen in a "natural" way so that the resulting system of maps indeed form an u*.m., that is, the naturality condition in the definition of "u*.m." coordinating the various instances is satisfied. The usual construction of the above diagram has to be replaced with something more elaborate.

The constructions to be given presently may look quite unmotivated, even bizarre, for the reader who approaches the larger subject of Stone duality for first order logic for the first time. Certainly, prior understanding of the papers [M1], [M2] would help a great deal in motivation, although there are genuinely new elements in the present context. We also face the recurring problem of presentation in logic in general that consists in the fact that one has to introduce *syntactical* concepts first whose "final cause" lies in their *semantical* use, and the latter is being clarified separately at a later stage only. The division made here of syntax and semantics is emphasized by the titles of this section and the next. I trust that the reader will in fact see the division as one of "syntax" and "semantics", despite the fact that these concepts are meant here in not quite the usual senses. A reading that takes sections 6 , 7 and 8 in reverse order (and also allows jumping in and out) could also help motivation.

Another remark concerning a technical point. The reader will sense that there is a certain redundancy in the five items (1) to (5) in the definition of cell-system (see (ii) below); more specifically, that items (2) and (3) look like simpler cases of items (4) and (5). Indeed, it would be possible to shorten that definition, and basically eliminate items (2) and (3); this would entail some corresponding savings later in this section too. However, this would force us to consider ultraproducts and ultralimits as special cases of a common generalization (in the *grounding* of cell-systems, see (iii), the items of type (2) and (3) are grounded to ultraproducts, those of type (4) and (5) to ultralimits). It is well-known that this can be done ("limit ultraproducts", see [K]). However, the fact remains that we need only ultraproducts and ω-type limit ultrapowers, and not the more sophisticated limit ultraproducts; the latter if present would play a purely administrative role (of cutting down the number five to three), which feels like an undesirable procedure to adopt.

Let (I, U) , (J, V) be ultrafilters. Then the subset $U\dot{\times}V$ of $\mathcal{P}(I\times J)$ defined by

$$X\in U\dot{\times}V \iff \{j\in J : \{i\in I: (i, j)\in X\} \in U\} \in V$$

($X\subset I\times J$) is an ultrafilter on $I\times J$. For a sequence $\vec{U} = \langle (I_1, U_1), \ldots, (I_n, U_n)\rangle$ of ultrafilters, the ultrafilter $\dot{\times}\vec{U} = U_1\dot{\times}\ldots\dot{\times}U_n = \dot{\times}_{1\le k\le n} U_k$ on the set $\prod_{1\le k\le n} I_k$ is defined (by induction) as $(U_1\dot{\times}\ldots\dot{\times}U_{n-1})\dot{\times}U_n$. Clearly, $U_1\dot{\times}\ldots\dot{\times}U_k =$ $(U_1\dot{\times}\ldots\dot{\times}U_n)\dot{\times}(U_{n+1}\dot{\times}\ldots\dot{\times}U_k)$ whenever $1\le n\le k<\omega$ (the empty product of ultrafilters is the trivial ultrafilter on the one-element set).

(i) Let Λ be a graph.

(a) A *truncated ultraproduct* in Λ is any tuple of the form $u = (\lambda, I, U, P, g)$, with $\lambda\in |\Lambda|$, (I, U) an ultrafilter, $P\in U$ and $g: P\longrightarrow |\Lambda|$.

(b) A *type*-4 *ultralimit* in Λ is any $\ell = (\lambda, n, \vec{U}, P, h)$ where $\lambda\in |\Lambda|$,

$n < \omega$, $\vec{U} = \langle (I_k, U_k) \rangle_{1 \le k < \omega}$ an ω-sequence of ultrafilters, $P \in \dot{\times}(\vec{U} \upharpoonright [1, n])$, and $h : P \longrightarrow |\Lambda|$.

(c) A *type-5 ultralimit* in Λ is any $\ell = (\lambda, n, \vec{U}, P, \vec{h})$ where $\lambda \in |\Lambda|$, $n < \omega$, $\vec{U} = \langle (I_k, U_k) \rangle_{1 \le k < \omega}$ an ω-sequence of ultrafilters, $P \in \dot{\times}(\vec{U} \upharpoonright [1, n])$, and $\vec{h} = \langle h_k \rangle_{n \le k < \omega}$ such that $h_k : P_k \longrightarrow |\Lambda|$ for all $n \le k < \omega$; we have used (and will use below) the abbreviation $P_k = P \times \prod_{n < j \le k} I_j$.

(ii) A *cell-system* Λ is a graph, also denoted by Λ, together with data as follows:

a well-founded partial ordering $\prec$ on the set $|\Lambda|$ of nodes of Λ;

a partition $|\Lambda| = \Lambda_0 \dot{\cup} \Lambda_1 \dot{\cup} \Lambda_2 \dot{\cup} \Lambda_3 \dot{\cup} \Lambda_4 \dot{\cup} \Lambda_5$,

disjoint sets $\mathcal{U}_2$ and $\mathcal{U}_3$ of truncated ultraproducts in Λ,

a set $\mathcal{L}_4$ of type-4 ultralimits, and a set $\mathcal{L}_5$ of type-5 ultralimits in Λ,

with these data satisfying the following conditions:

(1) For any $e \in \mathrm{Arr}(\Lambda)$, $\mathrm{codom}(e) \in \Lambda_1$; for every $\lambda \in \Lambda_1$ there is exactly one arrow with codomain λ, and for this $e : \mu \longrightarrow \lambda$ we have $\mu \prec \lambda$. *Notation*: $\mu = \mu_{(1)\lambda}$ [here, μ and the subscript (1) are invariable; the subscript λ is the variable] and $e = e_\lambda$.

(2) For any $u \in \mathcal{U}_2$, we have $\lambda_u \in \Lambda_2$; for every $\lambda \in \Lambda_2$ there is a unique $u \in \mathcal{U}_2$ with $\lambda_u = \lambda$, and for this u we have $g_u(i) \prec \lambda$ for all $i \in P_u$. *Notation*: $u = u_{(2)\lambda}$.

(3) For any $u \in \mathcal{U}_3$, we have $g_u(i) \in \Lambda_3$ for all $i \in P_u$. For any $\lambda \in \Lambda_3$, there are a unique $u \in \mathcal{U}_3$ and a unique $i \in P_u$ such that $\lambda = g_u(i)$; for this u we have $\lambda_u \prec \lambda$. *Notation*: $u = u_{(3)\lambda}$, $i = i_{(3)\lambda}$.

(4) For any $\ell \in \mathcal{L}_4$, we have $\lambda_\ell \in \Lambda_4$, and for every $\vec{i} \in P_\ell$, we have $h_\ell \vec{i} \prec \lambda_\ell$. For every $\lambda \in \Lambda_4$ there is a unique $\ell \in \mathcal{L}_4$ with $\lambda_\ell = \lambda$. *Notation*: $\ell = \ell_{(4)\lambda}$.

(5) For any $\ell \in \mathcal{L}_5$, we have $h_{k\ell}(\vec{i}) \in \Lambda_5$ and $\lambda_\ell \prec h_{k\ell}(\vec{i})$ for all $n_\ell \le k < \omega$ and all $\vec{i} \in P_{k\ell}$. For any $\lambda \in \Lambda_5$ there are a unique $\ell \in \mathcal{L}_5$ a unique k, $n_\ell \le k < \omega$, and a unique $\vec{i} \in P_{k\ell}$ such that $\lambda = h_{k\ell}(\vec{i})$. *Notation*: $\ell = \ell_{(5)\lambda}$.

If Λ_0 is a singleton, we say Λ is *rooted*; its *root*, λ_{in}, is the unique element of Λ_0. We may refer to the whole cell-system by the symbol Λ. We'll write $\Lambda_{0\Lambda}, \ldots,$ $\mathcal{U}_{2\Lambda}, \ldots$ for $\Lambda_0, \ldots \mathcal{U}_2, \ldots$ to show dependence on Λ if necessary.

(iii) A *grounded cell-system* $G:\Lambda \longrightarrow \Gamma$ is a cell-system Λ, with a $\overset{*}{u}$.g. Γ, a mapping ("grounding") $G:\Lambda \longrightarrow \Gamma$ of graphs and maps $\mathcal{U}_{2\Lambda} \cup \mathcal{U}_{3\Lambda} \longrightarrow \mathcal{U}_{\Gamma}$, $\mathcal{L}_{4\Lambda} \cup \mathcal{L}_{5\Lambda} \longrightarrow \mathcal{L}_{\Gamma}$, all denoted by the same letter G, such that:

if $u \in \mathcal{U}_{2\Lambda} \cup \mathcal{U}_{3\Lambda}$, then $G\lambda_u = \gamma_{Gu}$, $U_u = U_{Gu}$, $Gg_u(i) = g_{Gu}(i)$ for all $i \in P_u$;

if $\ell \in \mathcal{L}_{4\Lambda}$, then $G\lambda_\ell = \gamma_{G\ell}$, $U_{k\ell} = U_{k(G\ell)}$ $(1 \le k \le \omega)$, and $G(h_\ell \vec{i})$ is constant with value $(\gamma')_{G\ell}$;

if $\ell \in \mathcal{L}_{5\Lambda}$, then $G\lambda_\ell = \gamma_{G\ell}$, $U_{k\ell} = U_{k(G\ell)}$ $(1 \le k \le \omega)$, and $G(h_{k\ell} \vec{i})$ is constant with value $(\gamma')_{G\ell}$ for all $n \le k \le \omega$.

We may refer to the grounded cell-system $G:\Lambda \longrightarrow \Gamma$ by the symbol Λ alone.

(iv) A grounded cell-system $G:\Lambda \longrightarrow \Gamma$ is a *direct covering of* Γ if the following conditions are satisfied:

(1) If $\lambda \in |\Lambda|$, $e \in \mathrm{Arr}(\Gamma)$ and $G(\lambda) = \mathrm{dom}(e)$, then there is $\hat{e} \in \mathrm{Arr}(\Lambda)$ such that $G(\hat{e}) = e$.

(2) If $v \in \mathcal{U}_\Gamma$, $P \in U_v$ and $g: P \longrightarrow |\Lambda|$ such that $G \circ g = g_v \restriction P$, then there is $u \in \mathcal{U}_{2\Lambda}$ such that $P_u = P$, $G(u) = v$ and $g_u = g$.

(3) If $v \in \mathcal{U}_\Gamma$, $\lambda \in |\Lambda|$ and $G(\lambda) = \gamma_v$, then there is $u \in \mathcal{U}_{3\Lambda}$ such that $\lambda_u = \lambda$ and $G(u) = v$.

(4) If $m \in \mathcal{L}_\Gamma$, $n \in \omega$, $P \in \dot{\times}(\vec{U}_m \restriction [1, n])$ and $h: P \longrightarrow |\Lambda|$ such that $G(h(\vec{i})) = \gamma'_m$ for all $\vec{i} \in P$, then there is $\ell \in \mathcal{L}_{4\Lambda}$ such that $G(\ell) = m$, $n_\ell = n$, $P_\ell = P$ and $h_\ell = h$.

(5) If $m \in \mathcal{L}_\Gamma$, $\lambda \in |\Lambda|$ and $G(\lambda) = \gamma_m$, then there is $\ell \in \mathcal{L}_{5\Lambda}$ such that $\lambda_\ell = \lambda$ and $G(\ell) = m$.

(v lemma) For any $\overset{*}{u}$.g. Γ and $\gamma_0 \in |\Gamma|$, there is a direct covering $G:\Lambda \longrightarrow \Gamma$, with Λ a rooted cell-system, such that $G(\lambda_{in}) = \gamma_0$.

Proof. This is a rather straightforward transfinite construction. In a similar and slightly more complicated situation in (vii) below, a similar construction will be given in detail; the details here will be omitted.

(vi) (a) Let Λ be a cell-system. An *inverse covering* $F:\Sigma\longrightarrow\Lambda^{op}$ *of* Λ is a cell-system Σ together with a morphism of graphs $F:\Sigma\longrightarrow\Lambda^{op}$, and maps

$$\mathcal{U}_{2\Sigma}\longrightarrow\mathcal{U}_{3\Lambda},\ \mathcal{U}_{3\Sigma}\longrightarrow\mathcal{U}_{2\Lambda},\ \mathcal{L}_{4\Sigma}\longrightarrow\mathcal{L}_{5\Lambda},\ \mathcal{L}_{5\Sigma}\longrightarrow\mathcal{L}_{4\Lambda},$$

all denoted by the same symbol F, such that the following *compatibility* conditions are satisfied:

$u\in\mathcal{U}_{2\Sigma}\cup\mathcal{U}_{3\Sigma} \Longrightarrow F\sigma_u = \lambda_{Fu},\ U_u = U_{Fu},\ P_u\subset P_{Fu}$ and for all $i\in P_u$, we have $Fg_u i = g_{Fu} i$;

$\ell\in\mathcal{L}_{4\Sigma} \Longrightarrow F\sigma_\ell = \lambda_{F\ell},\ n_\ell\geq n_{F\ell},\ \vec{U}_\ell = \vec{U}_{F\ell},\ P_\ell\subset P_{n_\ell,F\ell},\ Fh_\ell(\vec{i}) = h_{n_\ell,F\ell}(\vec{i})\ \ (\vec{i}\in P_\ell)$;

$\ell\in\mathcal{L}_{5\Sigma} \Longrightarrow F\sigma_\ell = \lambda_{F\ell},\ n \underset{\text{def}}{=} n_\ell = n_{F\ell},\ \vec{U}_\ell = \vec{U}_{F\ell},\ P_\ell = P_{F\ell}$, and for all $n\leq k<\omega$ and all $\vec{i}\in P_k$, we have $Fh_{k\ell}(\vec{i}) = h_{F\ell}(\vec{i}\restriction[1,n])$;

and also, the following *covering* conditions are satisfied:

(1) If $\sigma\in|\Sigma|$, $F(\sigma)\in\Lambda_1$, then there is $e\in \mathrm{Arr}(\Sigma)$ with $\mathrm{dom}(e) = \sigma$ (and, hence, also $F(e) = e_{F\sigma}$, $F(\mathrm{codom}(e)) = \mu_{(1),F\sigma}$).

(2) If $v\in\mathcal{U}_{3\Lambda}$, $P\in U_v$, $P\subset P_v$ and $g:P\longrightarrow|\Sigma|$ such that $F\circ g = g_v\restriction P$, then there is $u\in\mathcal{U}_{2\Sigma}$ such that $P_u = P$, $F(u) = v$ and $g_u = g$.

(3) If $F(\sigma)\in\Lambda_2$, then there is $u\in\mathcal{U}_{3\Sigma}$ such that $\sigma_u = \sigma$ (and, hence, also $F(u) = u_{(2),F\sigma}$).

(4) If $m\in\mathcal{L}_{5\Lambda}$, $n\geq n_m$, $P\subset P_{nm}$, $P\in \dot{\times}(\vec{U}_m\restriction[1,n])$ and $h:P\longrightarrow|\Sigma|$ such that $F\circ h = h_{nm}\restriction P$, then there is $\ell\in\mathcal{L}_{4\Sigma}$ such that $F(\ell) = m$, $n_\ell = n$, $P_\ell = P$ and $h_\ell = h$.

(5) If $F(\sigma)\in\Lambda_4$, there is $\ell\in\mathcal{L}_{5\Sigma}$ such that $\sigma_\ell = \sigma$ (and, hence, also

$F(\ell) = \ell_{(4)}, F\sigma$).

(b) If $G:\Lambda \longrightarrow \Gamma$ is a grounded cell-system, $F:\Sigma \longrightarrow \Lambda^{op}$ is an inverse covering of cell-systems, "composing" G and F gives a grounding $GF:\Sigma \longrightarrow \Gamma^{op}$ of Σ ; here Γ^{op} is the u.$\overset{*}{\text{g}}$. whose underlying graph is Γ^{op} , the opposite of the graph Γ , and whose other specifications are the same as those of Γ .

(vii lemma) For any cell-system Λ , any set Σ_0 , and any mapping $F_0:\Sigma_0 \longrightarrow |\Lambda|$, there is an inverse covering $F:\Sigma \longrightarrow \Lambda^{op}$ such that $\Lambda_{0\Sigma} = \Sigma_0$, and $F\restriction\Sigma_0 = F_0$.

Proof. A *partial inverse covering* (pic) $F:\Sigma \longrightarrow \Lambda^{op}$ *of* Λ is defined as an inverse covering in (vi)(a), but with the covering conditions (1) to (5) dropped.

The cell-system Σ' *extends* the cell-system Σ (in notation: $\Sigma \subset \Sigma'$) if (i) Σ is an induced subgraph of Σ' ($|\Sigma| \subset |\Sigma'|$, and for σ , $\sigma' \in |\Sigma'|$, $\hom_{\Sigma}(\sigma,\sigma') = \hom_{\Sigma'}(\sigma,\sigma')$), (ii) $\Lambda_{0\Sigma'} = \Lambda_{0\Sigma}$, $\Lambda_{i\Sigma} = \Lambda_{i\Sigma'} \cap |\Sigma|$ $(1 \le i \le 5)$ and $\mathcal{U}_{i\Sigma} \subset \mathcal{U}_{i\Sigma'}$ $(i=2,3)$ [note that these condition mean something strong; as a consequence, if e.g. $u \in \mathcal{U}_{3\Sigma'}$ and $g_u(i) \in |\Sigma|$ even for one $i \in I_u$, then (since there must be $v \in \mathcal{U}_{3\Sigma}$ with $g_v(j) = g_u(i)$ for suitable j , this v must belong to $\mathcal{U}_{3\Sigma'}$, and finally, v must be equal to u) necessarily, $u \in \mathcal{U}_{3\Sigma}$], and (iii) $\prec_{\Sigma} \subset \prec_{\Sigma'}$ and if $\sigma_2 \prec_{\Sigma'} \sigma_1$ with $\sigma_1 \in |\Sigma|$, then $\sigma_2 \in |\Sigma|$ ($\prec_{\Sigma'}$ is an "end-extension" of $\prec_{\Sigma}$).

Let us fix Λ in what follows; a "pic" is a "pic of Λ ".

The pic $F':\Sigma' \longrightarrow \Lambda^{op}$ *extends* the pic $F:\Sigma \longrightarrow \Lambda^{op}$ (in notation $F \subset F'$) if $\Sigma \subset \Sigma'$ and each component-function of F' extends the corresponding component of F . If $\mathcal{F}$ is a family of pics in which the partial ordering $\subset$ of pics is directed, then the union $\bigcup \mathcal{F}$ defined in the obvious way is again a pic; the well-foundedness of the resulting $\prec$ is ensured by the "end-extension" condition built into $\subset$.

Let us say that the pic $F':\Sigma' \longrightarrow \Lambda^{op}$ *completes* the pic $F:\Sigma \longrightarrow \Lambda^{op}$ (in notation, $F \subset_c F'$) if $F \subset F'$, and the instances of the covering conditions for items coming from Σ are satisfied in Σ' . In detail, this means the following five conditions (1) to (5).

(1) If $\sigma \in |\Sigma|$ and $F(\sigma) \in \Lambda_1$, then there is $e \in \mathrm{Arr}(\Sigma')$ with $\mathrm{dom}(e) = \sigma$.

(2) If $v \in \mathcal{U}_{3\Lambda}$, $P \in U_v$, $P \subset P_v$ and $g:P \longrightarrow |\Sigma|$ such that $F \circ g = g_v \restriction P$, then there is $u \in \mathcal{U}_{2\Sigma'}$ such that $P_u = P$, $F'(u) = v$ and $g_u = g$.

(3) If $\sigma \in |\Sigma|$ and $F(\sigma) \in \Lambda_2$, then there is $u \in \mathcal{U}_{3\Sigma'}$ such that $\sigma_u = \sigma$.

(4) If $m \in \mathcal{L}_{5\Lambda}$, $n \geq n_m$, $P \subset P_{nm}$, $P \in \dot{\times}(\vec{U}_m \restriction [1, n])$ and $h: P \longrightarrow |\Sigma|$ such that $F \circ h = h_{nm} \restriction P$, then there is $\ell \in \mathcal{L}_{4\Sigma'}$ such that $F'(\ell) = m$, $n_\ell = n$, $P_\ell = P$ and $h_\ell = h$.

(5) If $\sigma \in |\Sigma|$ and $F(\sigma) \in \Lambda_4$, there is $\ell \in \mathcal{L}_{5\Sigma'}$ such that $\sigma_\ell = \sigma$.

We **claim** that for every pic F there is another pic F' such that $F \subset_c F'$. Once the claim is established, we prove the desired assertion as follows. Note that the given F_0 and Σ_0 form a pic $F_0: \Sigma_0 \longrightarrow \Lambda^{op}$; now, $\Lambda_{0\Sigma_0} = \Sigma_0$, $\Lambda_{i\Sigma_0} = \emptyset$ for $1 \leq i \leq 5$, $\mathcal{U}_{2\Sigma_0} = \mathcal{U}_{3\Sigma_0} = \mathcal{L}_{4\Sigma_0} = \mathcal{L}_{5\Sigma_0} = \emptyset$. By induction on the ordinal α, we define the pic $F_\alpha: \Sigma_\alpha \longrightarrow \Lambda^{op}$ by taking $F_{\alpha+1}$ to be one that completes F_α, and F_δ for δ a limit ordinal to be the union $\bigcup_{\alpha<\delta} F_\alpha$. Let κ be an infinite cardinal whose cofinality is greater than the cardinality of I_u for any $u \in \mathcal{U}_\Lambda$. Then $(F: \Sigma \longrightarrow \Lambda^{op}) = F_\kappa$ is the desired inverse covering of Λ. Indeed, we can check conditions (1) to (5) in the definition of "inverse covering" as follows. For (1), let $\sigma \in |\Sigma|$ such that $F(\sigma) \in \Lambda_{1\Lambda}$; there is $\alpha < \kappa$ such that $\sigma \in |\Sigma_\alpha|$; then $F_\alpha(\sigma) \in \Lambda_{1\Sigma_\alpha}$; hence, since $F_{\alpha+1}$ completes F_α, there is $e \in \mathrm{Arr}(\Sigma_{\alpha+1})$ with $\mathrm{dom}(e) = \sigma$; of course, $e \in \mathrm{Arr}(\Sigma)$, completing the verification for (1). For (2), let $v \in \mathcal{U}_{3\Lambda}$, $P \in U_v$, $P \subset P_v$ and $g: P \longrightarrow |\Lambda|$ such that $F \circ g = g_v$. By the choice of κ, $\mathrm{card}(P) < \mathrm{cf}(\kappa)$. It follows that there is $\alpha < \kappa$ such that $\mathrm{range}(g) \subset |\Sigma_\alpha|$. Then, by $F_\alpha \subset_c F_{\alpha+1}$, there is $u \in \mathcal{U}_{2\Sigma_{\alpha+1}} \subset \mathcal{U}_{2\Sigma}$ with the desired properties, completing the verification for (2). The rest of the conditions are similarly checked.

It remains to justify the claim above. Start with an arbitrary pic $F: \Sigma \longrightarrow \Lambda^{op}$; we construct $F': \Sigma' \longrightarrow \Lambda^{op}$ completing F in a straightforward manner; the complete description of the construction is lengthy only because of the number of items to be dealt with. $|\Sigma'|$ is the union of $|\Sigma|$ and the set of new nodes $\tau^{(-)}[-]$ indicated below; with each new node, the part $\Lambda_{i\Sigma'}$ of the partition of $|\Sigma'|$ in which the node is to be put, and the effect of F' on the node are also indicated. The set of arrows is extended by adding the new arrows $f^{(1)}[-]$, and similarly for the specification-sets $\mathcal{U}_2$, $\mathcal{U}_3$, $\mathcal{L}_4$, $\mathcal{L}_5$. The superscripts (1), (2), ... indicate the condition that requires the addition of the item in question. Here is the description of what is added to F:

$\tau^{(1)}[\sigma] \ (\in \Lambda_{1\Sigma'}) \overset{F'}{\longmapsto} \mu_{(1), F\sigma}$: one for each $\sigma \in |\Sigma|$ such that

$F(\sigma) \in \Lambda_1$;

$(f^{(1)}[\sigma] : \sigma \longrightarrow \tau^{(1)}[\sigma])$ $(\in \mathrm{Arr}(\Sigma'))$ $\overset{F'}{\longmapsto}$ $e_{F\sigma}$: one for each σ as above;

$\tau^{(2)}[g]$ $(\in \Lambda_{2\Sigma'})$ $\overset{F'}{\longmapsto}$ λ_v: one for each $v \in \mathcal{U}_{3\Lambda}$, $P \in U_v$ and $g: P \longrightarrow |\Sigma|$ such that $P \subset P_v$ and $F \circ g = g_v \restriction P$;

$u^{(2)}[g] = (\tau^{(2)}[g], I_v, U_v, P, g)$ $(\in \mathcal{U}_{3\Sigma'})$ $\overset{F'}{\longmapsto} v$: one for each v, P and g as above;

$\tau^{(3)}[\sigma, i]$ $(\in \Lambda_{3\Sigma'})$ $\overset{F'}{\longmapsto}$ $g_v(i)$: one for each $\sigma \in |\Sigma|$ and $v \in \mathcal{U}_{2\Lambda}$ such that $F(\sigma) = \lambda_v$ and for each $i \in P_v$;

$u^{(3)}[\sigma] = (\sigma, I_v, U_v, P_v, \langle i \in P_v \mapsto \tau^{(3)}[\sigma, i] \rangle)$ $(\in \mathcal{U}_{2\Sigma'})$ $\overset{F'}{\longmapsto} v$: one for each σ and v as above;

$\tau^{(4)}[h]$ $(\in \Lambda_{4\Sigma'})$ $\overset{F'}{\longmapsto} \lambda_m$: one for each $m \in \mathcal{L}_{5\Lambda}$, $n \geq n_m$, $P \in U_{nm}$, $P \in \dot{\times}(\vec{U}_m \restriction [1, n])$, and $h: P \longrightarrow |\Lambda|$ such that $P \subset P_{nm}$;

$\ell^{(4)}[h] = (\tau^{(4)}[h], n, \vec{U}_m, P, \langle \vec{i} \in P \mapsto F^{-1} h_{nm}(\vec{i}) \rangle)$ $(\in \mathcal{L}_{4\Sigma'})$ $\overset{F'}{\longmapsto} m$: one for each m, n, P and h as above;

$\tau^{(5)}[\sigma, \vec{i}]$ $(\in \Lambda_{5\Sigma'})$ $\overset{F'}{\longmapsto} h_m(\vec{i} \restriction [1, n_m])$: one for each $\sigma \in |\Sigma|$ and $m \in \mathcal{L}_{4\Lambda}$ such that $F(\sigma) = \lambda_m$, and for each $k \geq n_m$ and $\vec{i} \in P_{km}$;

$\ell^{(5)}[\sigma] = (\sigma, n_m, \vec{U}_m, P_m, \langle \vec{i} \in P_{km} \mapsto \tau^{(5)}[\sigma, \vec{i}] \rangle_{k \geq n_m}$ $(\in \mathcal{L}_{5\Sigma'})$ $\overset{F'}{\longmapsto} m$: one for each σ and m as above.

We define $\prec_{\Sigma'}$ to end-extend $\prec_\Sigma$: each new node is $\succ_{\Sigma'}$ every old node, and no two new nodes are comparable in $\prec_{\Sigma'}$; clearly, $\prec_{\Sigma'}$ is well-founded. Note that the compatibility conditions required for pics are met by F'; clearly, $F \subset_c F'$.

(viii) Let Λ be a cell-system. A *regular set in* Λ is a subset Ψ of $|\Lambda|$ satisfying the following conditions:

(0) $\Lambda_0 \subset \Psi$.

(1) If $\lambda \in \Lambda_1$, then $\lambda \in \Psi$ iff $\lambda_{(1)} \in \Psi$ (see (ii)(1)).

(2) If $\lambda \in \Lambda_2$, then $\lambda \in \Psi$ iff, for $u_{(2)\lambda} = (\lambda, I, U, P, g)$ (see (ii)(2)), we have that $\{i \in P: g(i) \in \Psi\} \in U$.

(3) For any $u = (\lambda, I, U, P, g) \in \mathcal{U}_3$, if $\lambda \in \Psi$, then $\{i \in P: g(i) \in \Psi\} \in U$. If $\lambda \in \Lambda_3 \cap \Psi$, then $\lambda_{u_{(3)\lambda}} \in \Psi$ (see (ii)(3)).

(4) If $\lambda \in \Lambda_4$, then $\lambda \in \Psi$ iff, for $\ell = \ell_{(4)\lambda}$ (see (ii)(4)), $\{\vec{i} \in P_\ell : h_\ell(\vec{i}) \in \Psi\} \in \dot{\times}(\vec{U}_\ell \restriction [1, n_\ell])$.

(5) For any $\ell = (\lambda, n, \vec{U}, P, \vec{h}) \in \mathcal{L}_5$, if $\lambda \in \Psi$, then there is $n \le n' < \omega$ such that for all $k \ge n'$ we have that $\{\vec{i} \in P_k : h_k(\vec{i}) \in \Psi\} \in \dot{\times}(\vec{U} \restriction [1, k])$. If $\lambda \in \Lambda_5 \cap \Psi$, then for $\ell = \ell_{(5)\lambda}$ (see (ii)(5)), we have $\lambda_\ell \in \Psi$.

$\mathrm{Reg}(\Lambda)$ denotes the set of regular sets of Λ. $\mathrm{Reg}(\Lambda)$ forms a meet semilattice with respect to set-theoretic containment as a partial order on $\mathrm{Reg}(\Lambda)$ (the proof is by a straightforward recursion along the well-founded $\prec$; compare (2), p. 135 in [M1]).

A *regular*$^-$ set is one that satisfies the conditions except possibly (0).

(ix) Let Λ be a cell-system. A *closed set* in Λ is a subset Ψ of $|\Lambda|$ satisfying one direction of the conditions (0) to (5) in (viii); that is, keep (0); replace "iff" by "if" in (1), (2) and (4); drop the second sentence in each of (3), (5). Call, for future reference, the so modified conditions (ix)(0) to (ix)(5).

Every closed set contains a regular set (defined by recursion along $\prec$).

$\mathrm{Cl}(\Lambda)$ is the set of closed sets of Λ.

A *closed*$^-$ set is one that satisfies the same conditions except possibly (0).

(x lemma) Let $F : \Sigma \longrightarrow \Lambda^{\mathrm{op}}$ be an inverse covering. If Φ is a closed$^-$ set in Σ, then $\Psi \underset{\mathrm{def}}{=} |\Lambda| - F''\Phi$ is a closed$^-$ set in Λ.

Proof. We verify the five conditions in "closed$^-$" for Ψ.

(1) Assume $\lambda \in \Lambda_1$ and $\mu_{(1)\lambda} \in \Psi$, to show $\lambda \in \Psi$. Suppose that, on the contrary, $\lambda \in F''\Phi$. We have $\sigma \in \Phi$, $F\sigma = \lambda$. By (vi)(a)(1), we have $e \in \mathrm{Arr}(\Sigma)$ with $\mathrm{dom}(e) = \sigma$. Then necessarily, $Fe = e_\lambda$; since Φ is closed$^-$, $\sigma' = \mathrm{codom}(e) \in \Phi$, and $\mu_{(1)\lambda} = \mathrm{dom}(e_\lambda) = F(\sigma') \in F''\Phi$, in contradiction to $\mu_{(1)\lambda} \in \Psi$.

(2) Let $\lambda \in \Lambda_2$; $u_{(2)\lambda} = (\lambda, I, U, P_0, g)$. Assume $P = \{i \in P_0 : gi \in \Psi\} \in U$; we need that $\lambda \notin F''\Phi$. Suppose, on the contrary, that $\lambda = F\sigma$, $\sigma \in \Phi$. Then by (vi)(a)(3), there is $v = (\sigma, I, U, P_1, h) \in \mathcal{U}_{3\Sigma}$ such that $P_1 \subset P_0$, $Fhi = gi$ $(i \in P_1)$. Since Φ is closed$^-$, $Q = \{i \in P_1 : hi \in \Phi\} \in U$. Since, by the definitions of Ψ and P, $gi \notin F''\Phi$ for all $i \in P$, we have $P \cap Q = 0$. Contradiction.

(3) Let $v = (\lambda, I, U, P_0, g) \in \mathcal{U}_{3\Lambda}$, $\lambda \in \Psi$; note that g is injective. Assume that $P = \{i \in P_0 : gi \in F''\Phi\} \in U$, to derive a contradiction; this will suffice. Choose for each $i \in P$ some $\sigma_i \in \Phi$ such that $F\sigma_i = gi$. Let $g' : P \longrightarrow |\Lambda|$ for which $g'(i) = \sigma_i$ $(i \in P)$; then $F \circ g' = g \upharpoonright P$. Choose $u = (\sigma, I, U, P, g') \in \mathcal{U}_{2\Sigma}$ according to (vi)(a)(2); then $g'i = \sigma_i \in \Phi$ for all $i \in P$. Since Φ is closed$^-$, we have $\sigma \in \Phi$, and of course, $F\sigma = \lambda$, in contradiction to $\lambda \in \Psi$.

(4) Let $\lambda \in \Lambda_4$, $m \underset{\text{def}}{=} \ell_{(4)}\lambda = (\lambda, n, \vec{U}, P_0, h)$. Assume $P = \{\vec{i} \in P_0 : h\vec{i} \in \Psi\} \in \dot{\times}(\vec{U} \upharpoonright [1, n])$. Suppose, contrary to the assertion, that $\lambda = F\sigma$, $\sigma \in \Phi$. By (vi)(a)(5), there is $\ell = (\sigma, n, \vec{U}, P^1, \vec{h})$ such that $F\ell = m$; thus, $Fh_k(i_1, \ldots, i_n, i_{n+1}, \ldots, i_k) = h(i_1, \ldots, i_n)$ for all $n \le k < \omega$ and $\langle i_1, \ldots i_k \rangle \in P^1_k$. Since Φ is closed$^-$, there is $n \le k < \omega$ such that

$$P^* \underset{\text{def}}{=} \{\vec{i} \in P^1_k : h_k\vec{i} \in \Phi\} \in \dot{\times}(\vec{U} | [1, k]) .$$

We have that

$$\{\vec{i}_0 \in \prod_{n<j\le k} I_j : \{\vec{i} \in \prod_{1\le j\le n} I_j : \vec{i}\wedge\vec{i}_0 \in P^*\} \in U_1 \dot{\times} \ldots \dot{\times} U_n\} \in U_{n+1} \dot{\times} \ldots \dot{\times} U_k .$$

Pick any element $\vec{i}_0$ in the last-displayed set, and let $P^+ = \{\vec{i} \in \prod_{1\le j\le n} I_j : \vec{i}\wedge\vec{i}_0 \in P^*\}$; we have $P^+ \in \dot{\times}(\vec{U} \upharpoonright [1, n])$. But, for all $\vec{i} \in P^+$, $h\vec{i} = Fh_k(\vec{i}\wedge\vec{i}_0) \in F''\Phi$, and thus, $P^+ \cap P = 0$, contradiction.

(5) Let $m = (\lambda, n, \vec{U}, P, \vec{h}) \in \mathcal{L}_{5\Lambda}$, and assume that $\lambda \in \Psi$. We will show that for all $n \le k < \omega$, $\{\vec{i} \in P_k : h_k\vec{i} \in \Psi\} \in \dot{\times}(\vec{U} \upharpoonright [1, k])$. Fix k with $n \le k < \omega$, and assume that $\{\vec{i} \in P_k : h_k\vec{i} \in \Psi\} \notin \dot{\times}(\vec{U} \upharpoonright [1, k])$, to derive a contradiction. Since $\dot{\times}(\vec{U} \upharpoonright [1, k])$ is an ultrafilter, we have that $P^* \underset{\text{def}}{=} \{\vec{i} \in P_k : h_k\vec{i} \in F''\Phi\} \in \dot{\times}(\vec{U} \upharpoonright [1, k])$. Pick $\sigma_{\vec{i}} \in \Phi$ such that $h_k\vec{i} = F\sigma_{\vec{i}}$ for $\vec{i} \in P^*$ and let $h : P \longrightarrow |\Lambda|$ for which $h(\vec{i}) = \sigma_{\vec{i}}$ $(\vec{i} \in P)$; $F \circ h = h_k \upharpoonright P$. Apply (vi)(a)(4) to get $\ell \in \mathcal{L}_{4\Lambda}$ such that $F\ell = m$, $n_\ell = k$, $P_\ell = P^*$, and $h_\ell = h$. Since Φ is closed$^-$, it follows that $\sigma_\ell \in \Phi$, hence, $\lambda = F\sigma_\ell \in F''\Phi$, contradiction.

(xi lemma) Let $F : \Sigma \longrightarrow \Lambda^{\text{op}}$ be an inverse covering of cell-systems. If Ψ is a regular$^-$ set in Λ, then $F^{-1}(\Psi)$ is a closed$^-$ set in Σ.

Proof. Assume Ψ regular$^-$ in Λ; $\Phi = F^{-1}(\Psi)$; we show the five conditions for "closed$^-$" for Φ. Actually, (1), (2) and (4) hold for Φ in the version for "regular"; however, the reverse conditions in (3) and (5) fail.

(1) Let $\sigma \in \Sigma_1$. Then the F-image of $e_\sigma : \sigma_{(1)} \to \sigma$, $Fe_\sigma : F\sigma \to F\sigma_{(1)}$ is the same as $e_\lambda : \lambda_{(1)} \to \lambda$ for $\lambda = F\sigma_{(1)}$. We have

$$\sigma \in \Phi \iff F\sigma \in \Psi \iff F\sigma_{(1)} \in \Psi \iff \sigma_{(1)} \in \Phi ,$$

the second equivalence since Ψ is regular$^-$.

(2) Let $u = (\sigma, I, U, P, g) \in \mathcal{U}_\Sigma$. Then $Fu = (F\sigma, I, U, P', g') \in \mathcal{U}_\Lambda$, $Fgi = g'i$ for $i \in P (\subset P')$.

(a) Assume that $\{i \in P : gi \in \Psi\} \in U$. Then, at least, $\{i \in P' : g'i \in \Phi\} \neq 0$, which, by the second part of (viii)(3) for Ψ, says that $F\sigma \in \Psi$, thus $\sigma \in \Phi$.

(b) Assume $\sigma \in \Phi$; that is, $F\sigma \in \Psi$. Hence, by the first part of (viii)(3) for Ψ, $\{i \in P' : g'i \in \Psi\} \in U$. Intersecting this set with the set $P \in U$, we get that $\{i \in P : Fgi \in \Psi\} \in U$. But $\{i \in P : gi \in \Phi\} = \{i \in P : Fgi \in \Psi\}$.

(3) for "closed": like (2)(b), but using the first part of (viii)(2) for Ψ.

(4) : like (2).

(5) Let $\ell = (\sigma, n, \vec{U}, P, \vec{h}) \in \mathcal{L}_{5\Sigma}$, assume $\sigma \in \Phi$. We will show that for any $n \leq k < \omega$, $P^* \underset{\text{def}}{=} \{\vec{i} \in P_k : h_k \vec{i} \in \Phi\} \in U^1$ for $U^1 \underset{\text{def}}{=} \dot{\times}(\vec{U}\restriction[1, k])$. Fix k. We have $F\ell = (F\sigma, n, \vec{U}, P, h)$ and $Fh_k(i_1, \ldots, i_n, i_{n+1}, \ldots i_k) = h(i_1, \ldots i_n)$. With $Q \underset{\text{def}}{=} \{\vec{i}' \in P : h\vec{i}' \in \Psi\}$, and the projection

$$\pi : I^1 \underset{\text{def}}{=} \prod_{1 \leq j \leq k} I_j \longrightarrow I^2 \underset{\text{def}}{=} \prod_{1 \leq j \leq n} I_j ,$$

it follows that $P^* = \pi^{-1}(Q)$. The "only if" part of (viii)(4) for Ψ says, since $F\sigma \in \Psi$, that $Q \in U^2 \underset{\text{def}}{=} \dot{\times}(\vec{U}\restriction[1, n])$. But $U^1 = U^2 \dot{\times} U^3$ for a suitable U^3, namely $U^3 = \dot{\times}(\vec{U}\restriction[n+1, k])$. It follows that $Q \in U^2 \Longrightarrow \pi^{-1}(Q) \in U^1$.

(xii) An *ultrascheme* is a quintuple $(G : \Lambda \to \Gamma, \varphi, T, W, \pi)$ where

$G : \Lambda \to \Gamma$ is a grounded rooted cell-system,

$\varphi:\mathrm{Reg}(\Lambda)\longrightarrow|\Lambda|$ such that $\varphi(\Psi)\in\Psi$ for all $\Psi\in\mathrm{Reg}(\Lambda)$,

W is an ultrafilter on T,

$\pi:T\longrightarrow\mathrm{Reg}(\Lambda)$, and

for any $\Psi\in\mathrm{Reg}(\Lambda)$, and with $[\Psi]\underset{\mathrm{def}}{=}\{\Phi\in\mathrm{Reg}(\Lambda):\Phi\subset\Psi\}$, we have $\pi^{-1}[\Psi]\in W$.

(xiii) Let $(G:\Lambda\to\Gamma,\ \varphi,\ T,\ W,\ \pi)$ be an ultrascheme. Let $Q\in W$ and let us select (by (vii)) an inverse covering $F:\Sigma\longrightarrow\Lambda^{\mathrm{op}}$ with $\Sigma_0=Q$, $F\restriction\Sigma_0=\varphi\pi\restriction Q$; call it $F^{(Q)}:\Sigma^{(Q)}\longrightarrow\Lambda^{\mathrm{op}}$. We consider $\Sigma^{(Q)}$ for each $Q\in W$; note that, for Q, $Q'\in W$, the elements of $Q\cap Q'$ are common to both $|\Sigma^{(Q)}|$ and $|\Sigma^{(Q')}|$; otherwise, let us keep the sets $|\Sigma^{(Q)}|$ and $|\Sigma^{(Q')}|$ disjoint.

Now, we describe a rooted cell-system Σ. σ_{in} is the node of a type-3 ultraproduct specification $(\sigma_{\mathrm{in}},\ T,\ W,\ T,\mathrm{id}_T)$; that is, the latter is put into $\mathcal{U}_{3\Sigma}$, and accordingly, each $t\in T$ is an element of Σ_3 $(=\Lambda_{3\Sigma})$.

Let Σ^- denote the union of all the $\Sigma^{(Q)}$, $Q\in W$, with the input nodes of the latter identified with the elements of T (and hence, occasionally with each other) as described. The graph Σ is the disjoint union of Σ^- and $\{\sigma_{\mathrm{in}}\}$.

Let us call F the mapping with domain Σ^- for which $F(t)=\varphi\pi(t)$ for $t\in T$, and which agrees with $F^{(Q)}$ on $\Sigma^{(Q)}$, for all $Q\in W$. F is a map of graphs of the form $F:\Sigma^-\longrightarrow\Lambda^{\mathrm{op}}$.

The grounding $H:\Sigma\longrightarrow\Gamma^{+\mathrm{op}}$ is as follows. Γ^+ is the ultragraph Γ, plus one new node γ^+, and one more ultraproduct $(\gamma^+,\ T,\ W,\ T, G\circ\varphi\circ\pi)$. $H(\sigma_{\mathrm{in}})=\gamma^+$, and on Σ^-, H is the composite of F and G.

The rest of the structure of Σ as a cell-system is inherited from the components $\Sigma^{(Q)}$ in the obvious way.

(xiv lemma) With the notation of (xiii), for every $\Phi\in\mathrm{Reg}(\Sigma)$ there is $\sigma\in\Phi$ such that $F(\sigma)=\lambda_{\mathrm{in}}$.

Proof. Let $\Phi\in\mathrm{Reg}(\Sigma)$; assume $\lambda_{\mathrm{in}}\notin F''\Phi$. Since Φ is regular, and $(\sigma_{\mathrm{in}},\ T,\ W,\ 1_T)\in\mathcal{U}_{3\Sigma}$, there is $Q\in W$ such that $Q\subset\Phi$. Consider the subset

$|\Sigma^{(Q)}|$ of $|\Sigma|$, and $\Phi' = \Phi \cap \Sigma^{(Q)}$; clearly, Φ' is a closed$^-$ set of $\Sigma^{(Q)}$. By (x), $\Psi' = |\Lambda| - F''\Phi'$ is a closed$^-$ set of Λ. Since by assumption, Ψ' also contains λ_{in}, Ψ' is closed; hence, there is a regular set Ψ of Λ contained in Ψ'; $\Psi \cap F''\Phi' = 0$. We have that $\pi^{-1}(\{\Xi \in \mathrm{Reg}\Lambda : \Xi \subset \Psi\}) \in W$; intersecting the last set with $Q \in W$, we obtain a non-empty set. Take any t in that intersection. Since $t \in Q$, $t \in \Phi'$. $Ft = \varphi\pi t \in \pi t \subset \Psi$; contradiction to $\Psi \cap F''\Phi' = 0$.

(xv) A *quasi-inverse* of an ultrascheme $S = (G:\Lambda \to \Gamma, \varphi, T, W, \pi)$ is any ultrascheme $(H:\Sigma \to \Gamma^{+\mathrm{op}}, \xi, T', W', \pi')$, where $H:\Sigma \to \Gamma^{+\mathrm{op}}$ is the grounded cell-system obtained from S as described in (xiii), equipped with the mapping F also described in (xiii), and such that $F(\xi(\Phi)) = \lambda_{\mathrm{in}}$ for every $\Phi \in \mathrm{Reg}(\Sigma)$ (see (xiv)).

(xvi lemma) With $(\Sigma, \ldots)$ a quasi-inverse of $(\Lambda, \ldots)$ via $F:\Sigma^- \longrightarrow \Lambda^{\mathrm{op}}$, we have

$$\Psi \in \mathrm{Reg}(\Lambda) \Longrightarrow F^{-1}(\Psi) \cup \{\sigma_{\mathrm{in}}\} \in \mathrm{Cl}(\Sigma).$$

Proof. Each closure condition with its hypothesis involving an element of Σ^- being in $F^{-1}(\Psi) \cup \{\sigma_{\mathrm{in}}\}$ is satisfied as a consequence of (xi). (0) is clearly satisfied. There remains an instance of (3), with $u = (\sigma_{\mathrm{in}}, T, W, T, 1_T)$. But $Q = \pi^{-1}\{\Xi \in \mathrm{Reg}(\Lambda) : \Xi \subset \Psi\} \in W$, and $\{\varphi\pi t : t \in Q\} \subset \Psi$, hence $Q \subset F^{-1}(\Psi)$, showing what we want.

(xvii) An *isomorphism ultrascheme* is a sequence $\vec{S} = \langle S_n \rangle_{n<\omega}$ of ultraschemes S_n such that S_{n+1} is a quasi-inverse of S_n for all $n < \omega$. An *isomorphism extension* of an ultrascheme S is any isomorphism ultrascheme $\vec{S}$ such that $S_0 = S$.

(xviii conclusion) Every ultrascheme can be extended to an isomorphism ultrascheme.

(xix lemma) The rooted ultraschemes $S = (\Lambda, \varphi, T, W, \pi)$, $S' = (\Lambda', \varphi', T', W', \pi')$ are said to be *paired* if the following are true:

the cell-systems Λ, Λ' are grounded on the same ultragraph: $G:\Lambda \longrightarrow \Gamma$, $G':\Lambda' \longrightarrow \Gamma$;

$G(\lambda_{\mathrm{in}}) = G'(\lambda_{\mathrm{in}})$, $T' = T$, $W' = W$, and $G \circ \varphi \circ \pi = G' \circ \varphi' \circ \pi'$.

If S and S' are paired, then there are quasi-inverses S^+, S'^+ to S, S', respectively, such that S^+, S'^+ are also paired.

Proof. Let Σ, Σ' be "quasi-inverse" cell-systems to S, S', respectively, given by (xiii), with connecting maps $F:\Sigma^- \longrightarrow \Lambda$, $F':\Sigma'^- \longrightarrow \Lambda'$, and let $\xi:\mathrm{Reg}(\Sigma) \longrightarrow |\Sigma|$, $\xi':\mathrm{Reg}(\Sigma') \longrightarrow |\Sigma'|$ be chosen by (xiv) so that $\xi(\Phi) \in \Phi$, $\xi(\Phi') \in \Phi'$, $F(\xi(\Phi)) = \lambda_{\mathrm{in}}$, $F'(\xi'(\Phi')) = \lambda'_{\mathrm{in}}$ ($\Phi \in \mathrm{Reg}(\Sigma)$, $\Phi' \in \mathrm{Reg}(\Sigma')$). Let $T = \mathrm{Reg}(\Sigma) \times \mathrm{Reg}(\Sigma')$, with $\rho: T \longrightarrow \mathrm{Reg}(\Sigma)$, $\rho': T \longrightarrow \mathrm{Reg}(\Sigma')$ the two projections. The system of sets

$$\mathcal{F} = \{\rho^{-1}[\Phi] : \Phi \in \mathrm{Reg}(\Sigma)\} \cup \{\rho'^{-1}[\Phi'] : \Phi' \in \mathrm{Reg}(\Sigma')\}$$

clearly has the finite intersection property (see the next to last sentence in (viii)); let W be any ultrafilter on T containing $\mathcal{F}$. Then, $(\Sigma, \xi, T, W, \rho)$, $(\Sigma', \xi', T, W, \rho')$ are quasi-inverses of S, S', respectively, and they are paired. Indeed, the fact that they are based on the ultragraph $\Gamma^+ = \Gamma'^+$ is ensured by the last condition of "paired" for S and S'; $H(\sigma_{\mathrm{in}}) = H'(\sigma'_{\mathrm{in}}) = \gamma^+$. The last condition for the new pair is obvious, because both $H \circ \xi \circ \rho$ and $H' \circ \xi' \circ \rho'$ are constant functions with value $G(\lambda_{\mathrm{in}}) = G(\lambda'_{\mathrm{in}})$.

(xx conclusion) Paired ultraschemes have paired iso-ultrascheme extensions (meaning that the components are pairwise paired).

7. THE SEMANTICS OF SPECIAL ULTRAMORPHISMS

With (I, U) , (J, V) ultrafilters , A a set, we have a bijection $\varphi = \varphi_A^{U,V} : (A^U)^V = A^{U,V} \xrightarrow{\cong} A^{U\dot{\times}V}$, defined by

$$\varphi(\langle\langle a_{i,j}\rangle_{i\in P_j}/U\rangle_{j\in P}/V) = \langle a_{i,j}\rangle_{j\in P,\ i\in P_j}/U\dot{\times}V \ .$$

Let $\vec{U} = \langle (I_1, U_1), \ldots, (I_n, U_n)\rangle$ be an n-sequence of ultrafilters. If A is a set, $A^{\vec{U}}$ denotes the iterated ultrapower $(\ldots((A^{U_1})^{U_2})\ldots)^{U_n}$. We have a canonical bijection

$$\varphi_A^{\vec{U}} = \varphi^{\vec{U}} : A^{\vec{U}} \xrightarrow{\ \cong\ } A^{\dot{\times}(\vec{U})}$$

defined by induction on n as the composite of

$$(A^{\vec{U}'})^{U_n} \xrightarrow[(\varphi_A^{\vec{U}'})^{U_n}]{} (A^{\dot{\times}(\vec{U}')})^{U_n} \xrightarrow[\varphi_A^{\dot{\times}(\vec{U}'),\, U_n}]{} A^{\dot{\times}(\vec{U})} \ .$$

If $P \in \dot{\times}(\vec{U})$, $f: P \longrightarrow A$, then $\langle f(\vec{i})\rangle_{\vec{i}\in P}/\vec{U}$ will denote the element $(\varphi_A^{\vec{U}})^{-1}(\langle f(\vec{i})\rangle_{\vec{i}\in P}/\dot{\times}(\vec{U}))$ of $A^{\vec{U}}$.

Usually, $\vec{U}$ will denote an ω-sequence of ultrafilters. In this case, if n is a positive integer, $P \in \dot{\times}(\vec{U}\restriction[1, n])$, $f: P \longrightarrow A$, then $\langle f(\vec{i})\rangle_{\vec{i}\in P}/\vec{U}$ will denote the element $\delta_A^{\vec{U},n}(\langle f(\vec{i})\rangle_{\vec{i}\in P}/\vec{U}\restriction[1, n])$, with the canonical map $\delta_A^{\vec{U},n} : A^{\vec{U}\restriction[1,n]} \longrightarrow A^{\vec{U}}$.

(i) (compare 6.7 in [M2]) Let $G: \Lambda \longrightarrow \Gamma$ be a rooted cell-system, $\mathcal{A}: \Gamma \longrightarrow \mathbf{Set}^*$ an ultra*diagram. A *filling* $\mathcal{F}$ *of* Λ *along* $\mathcal{A}$ is a function with domain a regular subset Ψ of Λ such that:

(0) For $\lambda \in \Psi$, we have $\mathcal{F}(\lambda) \in \mathcal{A}G\lambda$.

(1) For $e: \lambda \rightarrow \lambda' \in \mathrm{Arr}(\Lambda)$, if both λ , λ' are in Ψ , then $\mathcal{F}(\lambda') = (\mathcal{A}Ge)(\mathcal{F}\lambda)$.

(2/3) For $u = (\lambda, I, U, P_0, g) \in \mathcal{U}_{(2)\Lambda} \cup \mathcal{U}_{(3)\Lambda}$, if $\lambda \in \Psi$ (and hence, $P \underset{\text{def}}{=} \{i \in P_0 : gi \in \Psi\} \in U$), then $[\mathcal{A}, Gu](\mathcal{F}\lambda) = \langle \mathcal{F}gi \rangle_{i \in P} / U$.

(4) For $\ell = (\lambda, n, \vec{U}, P, h) \in \mathcal{L}_{4\Lambda}$, $G\ell = (\gamma, \gamma', \vec{U})$, if $\lambda \in \Psi$ (and hence, $P' = \{\vec{i} \in P : h\vec{i} \in \Psi\} \in \dot{\times}(\vec{U} \restriction [1, n])$), then we have

$$[\mathcal{A}, G\ell](\mathcal{F}(\lambda)) = \langle \mathcal{F}h(\vec{i}) \rangle_{\vec{i} \in P'} / \vec{U}.$$

(5) For $\ell = (\lambda, n, \vec{U}, P, \vec{h}) \in \mathcal{L}_{5\Lambda}$, $G\ell = (\gamma, \gamma', \vec{U})$, if $\lambda \in \Psi$, then for every k such that $P'_k \underset{\text{def}}{=} \{\vec{i} \in P_k : h\vec{i} \in \Psi\} \in \dot{\times}(\vec{U} \restriction [1, k])$ (almost all $k \in \omega$ are such as $\lambda \in \Psi$; see 6.(viii)(5)), we have

$$[\mathcal{A}, G\ell](\mathcal{F}(\lambda)) = \langle \mathcal{F}h_k(\vec{i}) \rangle_{\vec{i} \in P'_k} / \vec{U}.$$

(ii lemma) (compare 6.8 in [M2]) Let $G : \Lambda \longrightarrow \Gamma$ be a rooted cell-system, $\mathcal{A} : \Gamma \longrightarrow \mathbf{Set}^*$ an u.*d.

(a) Given any $a \in \mathcal{A}(G\lambda_{\mathrm{in}})$, there is at least one filling $\mathcal{F}$ of Λ along $\mathcal{A}$ such that $\mathcal{F}(\lambda_{\mathrm{in}}) = a$.

(b) If $\mathcal{F}$ and $\mathcal{G}$ are fillings of Λ along $\mathcal{A}$ with $\mathcal{F}(\lambda_{\mathrm{in}}) = \mathcal{G}(\lambda_{\mathrm{in}})$, then there is a regular set Ψ of Λ with $\Psi \subset \mathrm{dom}(\mathcal{F}) \cap \mathrm{dom}(\mathcal{G})$ such that $\mathcal{F} \restriction \Psi = \mathcal{G} \restriction \Psi$.

(c) More generally, if $\varphi : \mathcal{A} \longrightarrow \mathcal{B}$ is a lax morphism in $\mathrm{Hom}(\Gamma, \mathbf{Set})$, $\mathcal{F}$ and $\mathcal{G}$ are fillings of Λ along $\mathcal{A}$ and $\mathcal{B}$, respectively, with $\varphi_{G(\lambda_{in})}(\mathcal{F}(\lambda_{\mathrm{in}})) = \mathcal{G}(\lambda_{\mathrm{in}})$, then there is $\Psi \in \mathrm{Reg}(\Lambda)$ such that $\Psi \subset \mathrm{dom}(\mathcal{F}) \cap \mathrm{dom}(\mathcal{G})$ and $\varphi_{G\lambda}(\mathcal{F}\lambda) = \mathcal{G}\lambda$ for all $\lambda \in \Psi$.

The **proofs** are straightforward inductions along $\prec$; for similar, explicit, arguments, see (4)(*), p. 137 in [M1].

(iii lemma) Let $\mathcal{A} : \Gamma \longrightarrow \mathrm{Set}^*$ be an u.*d. Let $G : \Lambda \longrightarrow \Gamma$ be a direct covering of Γ. Then, for any filling $\mathcal{F}$ of the grounded cell-system $G : \Lambda \longrightarrow \Gamma$ along $\mathcal{A}$, the sets

$$\Sigma(\gamma) = \{\mathcal{F}(\lambda) : \lambda \in \mathrm{dom}(\mathcal{F}),\ G(\lambda) = \gamma\} \subset \mathcal{A}(\gamma)$$

determine a subobject $\Sigma_{\mathcal{F}}$ of $\mathcal{A}$ in $\mathrm{Hom}(\Gamma, \mathrm{Set}^*)$.

Proof. It is easily seen that a system $\Sigma = \langle \Sigma(\gamma) \rangle_{\gamma \in |\Gamma|}$ of subsets $\Sigma(\gamma) \subset \mathcal{A}(\gamma)$

determines a subobject of $\mathcal{A}$ in $\mathrm{Hom}(\Gamma, \mathrm{Set}^*)$ if and only if the following conditions (1) to (5) are satisfied (compare 4.5 in [M1] and 6.3 in [M2]):

(1) For any $(e:\gamma \longrightarrow \gamma') \in \mathrm{Arr}(\Gamma)$, if $x \in \Sigma(\gamma)$, then $\mathcal{A}(e)(x) \in \Sigma(\gamma')$.

(2) For any $v \in \mathcal{U}_\Gamma$, $P \in U_v$ and $x_i \in \Sigma(g_v(i))$ for $i \in P$, we have $[\mathcal{A}, v]^{-1}(\langle x_i \rangle_{i \in P}/U_v) \in \Sigma(\gamma_v)$.

(3) For any $v \in \mathcal{U}_\Gamma$ and $x \in \Sigma(\gamma_v)$, there are $P \in U_v$ and $x_i \in \Sigma(g_v(i))$ for all $i \in P$ such that $x = [\mathcal{A}, v]^{-1}(\langle x_i \rangle_{i \in P}/U_v)$.

(4) For any $m \in \mathcal{L}_\Gamma$, $n \in \omega$, $P \in \dot{\times}(\vec{U}_m \restriction [1, n])$ and $x_{\vec{i}} \in \Sigma(\gamma'_m)$ for all $\vec{i} \in P$, we have $[\mathcal{A}, m]^{-1}(\langle x_{\vec{i}} \rangle_{\vec{i} \in P}/U_{nm}) \in \Sigma(\gamma_m)$.

(5) For any $m \in \mathcal{L}_\Gamma$ and $x \in \Sigma(\gamma_m)$, there are $P \in \dot{\times}(\vec{U}_m \restriction [1, n])$ and $x_{\vec{i}} \in \Sigma(\gamma'_m)$ for all $\vec{i} \in P$ such that $x = [\mathcal{A}, m]^{-1}(\langle x_{\vec{i}} \rangle_{\vec{i} \in P}/U_{nm})$.

One verifies these conditions for $\Sigma = \Sigma_{\mathcal{F}}$ by using the corresponding conditions defining "direct covering"; see 6.(iv). E.g., to see (4), let $m \in \mathcal{L}_\Gamma$, $n \in \omega$, $P \in \dot{\times}(\vec{U}_m \restriction [1, n])$ and $x_{\vec{i}} \in \Sigma(\gamma'_m)$ for all $\vec{i} \in P$. By definition of $\Sigma(\gamma'_m)$, for each $\vec{i} \in P$ there is $\lambda_{\vec{i}} \in \mathrm{dom}(\mathcal{F})$ such that $G(\lambda_{\vec{i}}) = \gamma'_m$ and $x_{\vec{i}} = \mathcal{F}(\lambda_{\vec{i}})$. Let $h(\vec{i}) = \lambda_{\vec{i}}$ $(\vec{i} \in P)$. By 6.(iv)(4), there is $\ell = (\lambda, I_v, U_v, n, P, h) \in \mathcal{L}_{4\Lambda}$; since $\mathrm{dom}(\mathcal{F})$ is a regular set, $\lambda \in \mathrm{dom}(\mathcal{F})$. Also, $\mathcal{F}(\lambda) = [\mathcal{A}, m]^{-1}(\langle x_{\vec{i}} \rangle_{\vec{i} \in P}/U_{nm})$, by the definition of "filling". Since $\mathcal{F}(\lambda) \in \Sigma(\gamma_m)$, (4) is verified.

(iv) (a) Let $S = (G:\Lambda \to \Gamma, \varphi, T, W, \pi)$ be an ultrascheme (see 6.(xii)). Let $\mathcal{A}:\Gamma \longrightarrow \mathbf{Set}$ be an u.d. We let $D^S_{\mathcal{A}} = \mathcal{A}(G\lambda_{\mathrm{in}})$, $C^S_{\mathcal{A}} = \prod_{t \in T} \mathcal{A}G\varphi(\pi t)/W$, and we define the function

$$\mu = \mu^S_{\mathcal{A}} : D^S_{\mathcal{A}} \longrightarrow C^S_{\mathcal{A}} \tag{1}$$

as follows. We let $\mathcal{F}$ be any filling of Λ along $\mathcal{A}$ such that $\mathcal{F}(\lambda_{\mathrm{in}}) = a$, and let

$$\mu(a) = \langle \mathcal{F}(\varphi(\pi t)) \rangle_{t \in \pi^{-1}[\mathrm{dom}(\mathcal{F})]}/W. \tag{1'}$$

Recall that $[\Psi] = \{\Xi \in \mathrm{Reg}(\Lambda) : \Xi \subset \Psi\}$, that $\mathrm{dom}(\mathcal{F}) \in \mathrm{Reg}(\Lambda)$, and that for all $\Psi \in \mathrm{Reg}(\Lambda)$, $\pi^{-1}[\Psi] \in W$.

By (ii)(a) and (b), μ is well-defined.

(b) Moreover, let Γ^+ be the $\overset{*}{\text{u.g.}}$ which is obtained by adding one new node γ^+ to Γ, with one ultraproduct specification $u^+ = (\gamma^+, T, W, G\varphi\pi)$; note that Γ^+ is the same $\overset{*}{\text{u.g.}}$ as the one with the same name in 6.(xiii). For $\mathcal{A}^+ \in \mathrm{Hom}(\Gamma^+, \mathrm{Set})$, let $\mu^S_{\mathcal{A}^+} : \mathcal{A}^+(G\lambda_{\mathrm{in}}) \longrightarrow \mathcal{A}^+(\gamma^+)$ be defined by $\mu^S_{\mathcal{A}^+}(a) = [\mathcal{A}^+, u^+]^{-1}(\mu^S_{\mathcal{A}^+\restriction\Gamma}(a))$ for all $a \in \mathcal{A}^+(G\lambda_{\mathrm{in}})$, with $\mu^S_{\mathcal{A}^+\restriction\Gamma}$ from (1) and (1').

We claim that

$$\mu^S = \langle \mu^S_{\mathcal{A}^+} \rangle_{\mathcal{A}^+ \in \mathrm{Hom}(\Gamma^+, \mathbf{Set})} \text{ is an ultramorphism of type } (\Gamma^+, G\lambda_{\mathrm{in}}, \gamma^+)$$

in Set.

This follows from (ii)(c).

(v) Let S be as in (iv), $S' = (H:\Sigma \longrightarrow \Gamma^{+\mathrm{op}}, \xi, T', W', \pi')$ a quasi-inverse of S (see 6.(xv)). Recall the morphism $F:\Sigma^- \longrightarrow \Lambda^{\mathrm{op}}$.

Let $\mathcal{A}^+ : \Gamma^+ \longrightarrow \mathbf{Set}^*$ be an $\overset{*}{\text{u.d.}}$; we assume (for simplicity) that $\mathcal{A}^+$ is strict with respect to u^+, i.e., $[\mathcal{A}^+, u^+]$ is an identity. We then have the $\overset{*}{\text{u.d.}}$ $(\mathcal{A}^+)^{-1} : \Gamma^{+\mathrm{op}} \longrightarrow \mathbf{Set}^*$ which is the same as $\mathcal{A}^+$ except that on arrows $e:\gamma \longrightarrow \gamma'$ in Γ, that is, on arrows $e:\gamma' \longrightarrow \gamma$ in Γ^{op}, we put $(\mathcal{A}^+)^{-1}(e) = \mathcal{A}^+(e)^{-1}$ (all arrows in $\mathcal{A}^+$ are invertible). Let $\mathcal{A} = \mathcal{A}^+\restriction\Gamma$.

(a) Let $\mathcal{F}$ be a filling of Λ along $\mathcal{A}$. By 6.(xvi), let Φ be a regular set in Σ such that $\Phi \subset F^{-1}(\mathrm{dom}(\mathcal{F})) \cup \{\sigma_{\mathrm{in}}\}$. Then $\mathcal{G}$ defined by

$$\mathrm{dom}(\mathcal{G}) = \Phi,$$

$$\mathcal{G}(\sigma) = \mathcal{F}F(\sigma) \quad (\sigma \in F^{-1}(\mathrm{dom}(\mathcal{F}))),$$

$$\mathcal{G}(\sigma_{\mathrm{in}}) = \langle \mathcal{F}(\varphi(\pi t)) \rangle_{t \in \pi^{-1}\mathrm{dom}(\mathcal{F})} / W = \mu^S_{\mathcal{A}^+}(\mathcal{F}\lambda_{\mathrm{in}}) \quad \text{(see (1'))}$$

is a filling of Σ along $(\mathcal{A}^+)^{-1}$.

Proof. Essentially, by inspection. For clause (i)(5) for $\mathcal{G}$, we use the identity

$$\langle a_{\vec{i}\restriction[1,n]}\rangle_{\vec{i}\in P}/\vec{U} = \langle a_{\vec{i}'}\rangle_{\vec{i}'\in\pi''P}/\vec{U};$$

here, $n \le k < \omega$, $\pi\colon \prod_{1\le j\le k} I_j \longrightarrow \prod_{1\le j\le n} I_j$ is the projection, $P \in \dot{\times}(\vec{U}\restriction[1,k])$, and $\langle a_{\vec{i}'}\rangle_{\vec{i}'\in\pi''P}$ is a family of elements of the set A.

(b) By the definition of "quasi-inverse", $H\sigma_{\mathrm{in}} = \gamma^+$, $F\xi\pi' t$ is constant λ_{in} $(t \in T')$, hence, $H\xi\pi' t$ is constant $G\lambda_{\mathrm{in}}$. By (iii) applied to S and S', we have the arrows

$$\mu^S_{\mathcal{A}^+} \quad : \quad \mathcal{A}^+(G\lambda_{\mathrm{in}}) \longrightarrow \mathcal{A}^+(\gamma^+),$$

$$\mu^{S'}_{(\mathcal{A}^+)^{-1}} : \quad \mathcal{A}^+(\gamma^+) \longrightarrow \mathcal{A}^+(G\lambda_{\mathrm{in}})^{W'}.$$

Then the triangle

$$\begin{array}{ccc} \mathcal{A}^+(G\lambda_{\mathrm{in}}) & \xrightarrow{\quad\mu^S_{\mathcal{A}^+}\quad} & \mathcal{A}^+(\gamma^+) \\ \delta^{W'}_{G\lambda_{\mathrm{in}}}\Big\downarrow & \swarrow\ \mu^{S'}_{(\mathcal{A}^+)^{-1}} & \\ \mathcal{A}^+(G\lambda_{\mathrm{in}})^{W'} & & \end{array}$$

commutes.

Proof. Take any $a \in \mathcal{A}(G\lambda_{\mathrm{in}})$, let $\mathcal{F}$ be any filling of Λ along $\mathcal{A}$ with $\mathcal{F}(\lambda_{\mathrm{in}}) = a$. Consider $\mathcal{G}$ as in (a). We have that $\mathcal{G}(\sigma_{\mathrm{in}}) = \mu^S_{\mathcal{A}}(a)$. Thus, according to the definition of $c \underset{\mathrm{def}}{=} \mu^{S'}_{(\mathcal{A}^+)^{-1}}(\mu^S_{\mathcal{A}}(a))$, c is equal to $\langle \mathcal{G}\xi\pi' t\rangle_{t\in\pi'^{-1}[\Phi]}/W'$. But $\mathcal{G}\xi\pi' t = \mathcal{F}F\xi\pi' t = \mathcal{F}\lambda_{\mathrm{in}} = a$ for all $t \in \pi'^{-1}[\Phi]$. Thus, $c = \langle a\rangle/W' = \delta^{W'}(a)$ as desired.

(vi) Let $\vec{S} = \langle S_n\rangle_{n<\omega}$ be an isomorphism ultrascheme. $\vec{S}$ gives rise to a full ultra*morphism $\nu^{\vec{S}}$ in $\mathbf{Set}^*$ as follows. The following schematic picture should help following the construction:

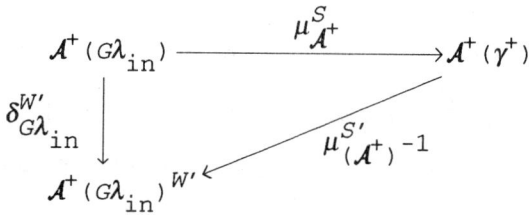

$$\begin{array}{ccc}
G_0\lambda_{\mathrm{in}} = \gamma_0 & \xrightarrow{\quad \mu^{S_0} \quad} & \gamma_1 \\
\delta^{W_0}\big\downarrow & \swarrow{\scriptstyle \mu^{S_1}} & \big\downarrow\delta^{W_1} \\
\gamma_2 & \xrightarrow{\quad \mu^{S_2} \quad} & \gamma_3 \\
\delta^{W_2}\big\downarrow & \swarrow{\scriptstyle \mu^{S_3}} & \big\downarrow\delta^{W_3} \\
\gamma_4 & \xrightarrow{\quad \mu^{S_4} \quad} & \gamma_5 \\
\vdots & & \vdots \\
\gamma_\omega & \overset{\cong}{\dashrightarrow} & \gamma_\omega^+
\end{array} .$$

Let $S_n = \langle G_n : \Lambda_n \to \Gamma_n, \varphi_n, T_n, W_n, \pi_n \rangle$; when $n = 0$, let us omit the subscript n . Note that Γ_n has one more node, denoted γ_n , with respect to Γ_{n-1} $(n > 0)$; in fact, $\gamma_n = G_n((\lambda_{\mathrm{in}})_{\Lambda_n})$. Let $\vec{W} = \langle W_{2n-1} \rangle_{1 \le n < \omega}$ and $\vec{W}' = \langle W_{2n} \rangle_{1 \le n < \omega}$.

We let Γ_ω be the $\mathrm{u}^*\mathrm{g}$. obtained from $\Gamma = \Gamma_0$ by adding three nodes and three specifications: the node γ^+ , the ultraproduct $u^+ = (\gamma^+, T, W, G\varphi\pi)$ (thus, Γ_ω contains Γ^+ , as defined in (iv)(b)), the further nodes γ_ω, γ'_ω, and the ultralimit specifications $\ell_\omega = (\gamma_\omega, \gamma_0, \vec{W})$ and $\ell'_\omega = (\gamma'_\omega, \gamma_1, \vec{W}')$. The $\mathrm{u}^*\mathrm{m}$. $v^{\vec{S}}$ will be of type $(\Gamma_\omega, \gamma_\omega, \gamma^+_\omega)$.

We build the auxiliary $\mathrm{u}^*\mathrm{g}$. Γ^ω by taking $\bigcup_{n<\omega} \Gamma_n$ and adding the new nodes γ_ω, γ'_ω. That is, Γ^ω has the additional nodes γ_n $(1 \le n < \omega)$, γ_ω, γ'_ω with respect to Γ_0 ; it also has the additional ultraproduct specifications $u_n = (\gamma_n, T_{n-1}, W_{n-1}, G_{n-1}\varphi_{n-1}\pi_{n-1})$ $(1 \le n < \omega)$, each of which except u_1 is an ultrapower, and ℓ_ω, ℓ'_ω defined above. $\gamma^+ = \gamma_1$ and $u^+ = u_1$. Γ_ω is a sub-$\mathrm{u}^*\mathrm{g}$. of Γ^ω .

The ultramorphism μ^{S_n} is defined for arguments that are ultra*diagrams of type Γ^+_n . For even n , Γ^+_n is a sub-ultra*graph of Γ^ω ; in fact, it has, beyond Γ_0 , the γ_k for $k \le n+1$, with the corresponding ultraproduct specifications. For odd n , Γ^+_n is a sub-ultra*graph of $\Gamma^{\omega\mathrm{op}}$.

Let $\mathcal{A}_\omega : \Gamma_\omega \to \mathbf{Set}^*$, and assume first that $\mathcal{A}_\omega$ is strict with respect the three specifications u^+ , ℓ_ω, ℓ'_ω ; that is, the corresponding transition isomorphisms $[\mathcal{A}, \gamma^+]$, $[\mathcal{A}, \ell_\omega]$, $[\mathcal{A}, \ell'_\omega]$ are identities. Let $\mathcal{A}^\omega$ be the unique strict $\mathrm{u}^*\mathrm{d}$. $\mathcal{A}^\omega : \Gamma^\omega \to \mathbf{Set}^*$ extending $\mathcal{A}_\omega$. For even n , let $\mathcal{A}_n = \mathcal{A}^\omega \restriction \Gamma^+_n$; for odd n , $\mathcal{A}_n = (\mathcal{A}^\omega \restriction (\Gamma^+_n))^{-1}$, where, as in (v), the last notation means taking the inverses of the isomorphisms involved. Then, by (iv) applied to each S_n , we have the functions

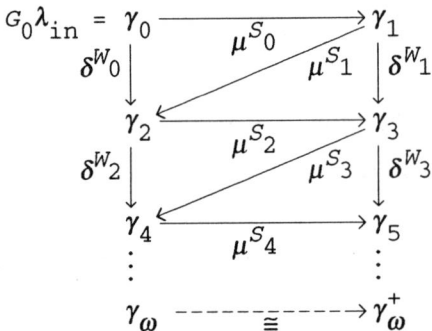

$$\mu_n \underset{\mathrm{def}}{=} \mu^{S_n}_{\mathcal{A}_n} : \mathcal{A}^{\omega}(\gamma_n) \longrightarrow \mathcal{A}^{\omega}(\gamma_{n+1}) \, .$$

Also, note that $\mathcal{A}^{\omega}(\gamma_{n+2}) = (\mathcal{A}^{\omega}(\gamma_n))^{W_{n+1}}$, and we have the diagonal maps

$$\delta_n = \delta^{W_{n+1}}_{\mathcal{A}^{\omega}\gamma_n} : \mathcal{A}^{\omega}(\gamma_n) \longrightarrow \mathcal{A}^{\omega}(\gamma_{n+2}) \, .$$

Moreover, we have that $\mathcal{A}^{\omega}(\gamma_{\omega}) = \mathcal{A}^{\omega}(\gamma_0)^{\vec{W}}$ and $\mathcal{A}^{\omega}(\gamma^{+}_{\omega}) = \mathcal{A}^{\omega}(\gamma_1)^{\vec{W}'}$. By (v)(b), the diagram

$$\begin{array}{ccccc}
\mathcal{A}^{\omega}\gamma_0 & & \xrightarrow{\ \mu_0\ } & & \mathcal{A}^{\omega}\gamma_1 \\
{\scriptstyle \delta_0}\downarrow & & \swarrow{\scriptstyle \mu_1} & & \downarrow{\scriptstyle \delta_1} \\
\mathcal{A}^{\omega}\gamma_2 & & \xrightarrow{\ \mu_2\ } & & \mathcal{A}^{\omega}\gamma_3 \\
{\scriptstyle \delta_2}\downarrow & & \swarrow{\scriptstyle \mu_3} & & \downarrow{\scriptstyle \delta_3} \\
\mathcal{A}^{\omega}\gamma_4 & & \xrightarrow{\ \mu_4\ } & & \mathcal{A}^{\omega}\gamma_5 \\
\vdots & & & & \vdots
\end{array}$$

commutes. Now, we may define the function

$$\nu = \nu^{\vec{S}}_{\mathcal{A}_{\omega}} : \mathcal{A}_{\omega}(\gamma_{\omega}) \longrightarrow \mathcal{A}_{\omega}(\gamma^{+}_{\omega})$$

by the prescription

$$\nu(\delta^{\vec{W},\,n}(a)) = \delta^{\vec{W}',\,n}(\mu_{2n}(a)) \quad \text{whenever } a \in \mathcal{A}^{\omega}(\gamma_{2n}) \, ;$$

in other words, ν is defined so that

$$\begin{array}{ccc}
\mathcal{A}^{\omega}\gamma_{2n} & \xrightarrow{\ \mu_{2n}\ } & \mathcal{A}^{\omega}\gamma_{2n+1} \\
{\scriptstyle \delta^{\vec{W},\,n}}\downarrow & & \downarrow{\scriptstyle \delta^{\vec{W}',\,n}} \\
\mathcal{A}^{\omega}\gamma_{\omega} & \xrightarrow[\ \nu\ ]{} & \mathcal{A}^{\omega}\gamma^{+}_{\omega}
\end{array}$$

commutes, where the vertical arrows are colimit coprojections. ν is indeed well-defined, and it is a bijection, its inverse being given by

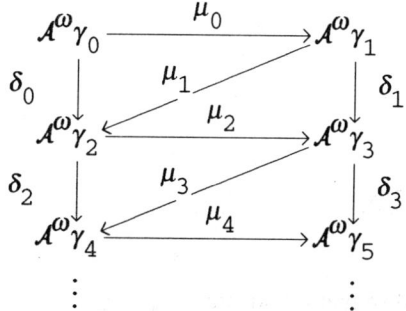

$$\nu^{-1}(\delta^{\vec{W}',n}(b)) = \delta^{\vec{W},n+1}(\mu_{2n+1}(b)) \qquad \text{whenever } b\in \mathcal{A}^{\omega}(\gamma_{2n+1}) .$$

We have defined the component $\nu^{\vec{S}}_{\mathcal{A}_\omega}$ of the $\nu^{\vec{S}}$ at $\mathtt{u}^*\mathtt{d}$.'s $\mathcal{A}_\omega$ that are strict with respect to u^+, ℓ_ω, ℓ'_ω.

For a general $\mathcal{B}_\omega:\Gamma_\omega\longrightarrow\mathbf{Set}$, let $\mathcal{A}=\mathcal{B}\upharpoonright\Gamma$, let $\mathcal{A}_\omega$ be the unique extension of $\mathcal{B}_\omega\upharpoonright\Gamma$ to Γ_ω which is strict with respect to u^+, ℓ_ω, ℓ'_ω. We have a unique isomorphism $\zeta:\mathcal{B}_\omega\xrightarrow{\cong}\mathcal{A}_\omega$ extending the identity on Γ; the three other components are

$$\zeta_{\gamma_1} = [\mathcal{B}_\omega,\gamma_1] \qquad : \mathcal{B}_\omega\gamma_1\xrightarrow{\ \cong\ }\mathcal{A}_\omega\gamma_1 ,$$

$$\zeta_{\gamma_\omega} = [\mathcal{B}_\omega,\gamma_\omega] \qquad : \mathcal{B}_\omega\gamma_\omega\xrightarrow{\ \cong\ }\mathcal{A}_\omega\gamma_\omega ,$$

and

$$\zeta_{\gamma'_\omega} = [\mathcal{B}_\omega,\gamma_1]^{\vec{W}'}\circ[\mathcal{B}_\omega,\gamma'_\omega] \ : \mathcal{B}_\omega\gamma'_\omega\xrightarrow{\ \cong\ }\mathcal{A}_\omega\gamma'_\omega .$$

The component $\nu^{\vec{S}}_{\mathcal{B}}$ has to be, and is, defined so that the square

$$\begin{array}{ccc}
\mathcal{A}_\omega\gamma_\omega & \xrightarrow{\ \nu^{\vec{S}}_{\mathcal{A}}\ } & \mathcal{A}_\omega\gamma'_\omega \\
\uparrow{\scriptstyle \zeta_{\gamma_\omega}} & & \uparrow{\scriptstyle \zeta_{\gamma'_\omega}} \\
\mathcal{B}_\omega\gamma_\omega & \xrightarrow[\ \nu^{\vec{S}}_{\mathcal{B}}\ ]{} & \mathcal{B}_\omega\gamma'_\omega
\end{array}$$

commutes.

$\nu=\langle\nu^{\vec{S}}_{\mathcal{A}_\omega}\rangle_{\mathcal{A}_\omega\in\mathrm{Hom}(\Gamma_\omega,\mathbf{Set})}$ so defined is a full ultra*morphism $\nu^{\vec{S}}$ in $\mathbf{Set}$ of type $(\Gamma_\omega,\gamma_\omega,\gamma^+_\omega)$. This is easy to see, by using the fact that the μ^{S_n}'s are ultramorphisms. In fact, first one deals with the naturality condition only for $\mathtt{u}^*\mathtt{d}$.'s $\mathcal{A}_\omega$ satisfying the strictness condition above; the general case will follow.

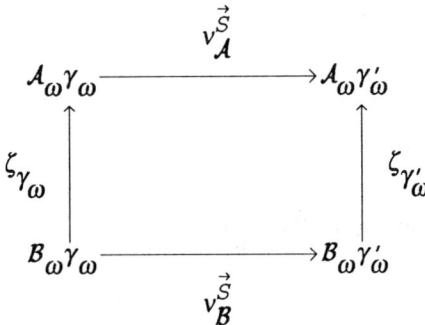

(vii) The *special* ultra^*morphisms on $\mathbf{Set}$ are the ones of the form $v^{\vec{S}}$ with an isomorphism ultrascheme $\vec{S}$. The special ultramorphisms on $\mathbf{Hom}^*(\mathcal{T}, \mathcal{S}et)$ are the ones induced by the special $\text{u}^*\text{m.}$'s on $\mathbf{Set}^*$. In other words, we take Δ to be the class of special (full) $\text{u}^*\text{m.}$'s on $\mathbf{Set}$, and make the constructions of Section 5 with this Δ in mind.

8. THE DUALITY THEOREM

The class Δ of full ultra*morphisms in $\mathbf{Set}^*$, kept arbitrary in Section 5, is now taken to be the class of special ultra*morphisms, introduced in the previous section. With any pretopos T, also written as $\mathcal{T}$ for emphasis, we have the pretopos morphism

$$\varepsilon_T = \varepsilon_{\mathcal{T}} : \mathcal{T} \longrightarrow \mathcal{H}om(\mathbf{Hom}^*(\mathcal{T}, \mathcal{S}et), \mathbf{Set})$$

introduced in 5.(vi)(b). $\mathbf{Hom}^*(\mathcal{T}, \mathcal{S}et)$ is the groupoid of models of T, equipped with the ultraproducts, ultralimits, and special ultramorphisms inherited from the category of sets. We also write $\mathbf{Mod}^* T$ for $\mathbf{Hom}^*(\mathcal{T}, \mathcal{S}et)$. $\mathcal{H}om(\mathbf{Mod}^* T, \mathbf{Set})$ is the Boolean pretopos of all structure preserving maps from $\mathbf{Mod}^* T$ to $\mathbf{Set}$; more precisely, the category of all Δ-ultra*functors, and lax ultra*transformations, with Δ as the class of special ultramorphisms.

1. Theorem. For any small Boolean pretopos T, e_T is an equivalence of categories.

Since the statement of the theorem involves the notion of special ultra*morphism, it cannot be said to have an immediately comprehensible content. However, the theorem directly entails the weakened version in which Δ is replaced by the class of all full ultra*morphisms on Set. In this version, the theorem does express a conceptual fact about the algebraic aspects of the category of sets endowed with the Boolean pretopos structure on the one hand, and the ultraproduct/ultralimit structure on the other. The main result of [M1] was the corresponding result for (general) pretoposes.

We do not have a proof of the general/weaker version of the theorem that is independent of the treatment in this paper. On the other hand, we do need Theorem 1. in its present formulation, since we do not have the generalized version of the preservation result 10.(viii) below, with general u.*m.'s in place of special ones.

The still stronger result to be stated next is the one that we will really need. We now refer to the context of the Stone-type adjunction

$$\mathcal{BP}^{*\mathrm{op}} \underset{F\ =\ \mathcal{H}om(-,\ \mathcal{S}et)}{\overset{G\ =\ \mathbf{Hom}^*(-,\ \mathcal{S}et)}{\rightleftarrows}} \mathbf{UG}\,.$$

of Section 5.

2. Theorem. For any small Boolean pretopos T, $\boldsymbol{K} \in \mathbf{UG}$ and $Z{:}\boldsymbol{K} \longrightarrow \mathbf{Mod}^* T$ in $\mathbf{UG}$, if Z is full, faithful and relation-conservative, then its transpose in the Stone adjunction, $Z^\# : T \longrightarrow \mathcal{H}om(\boldsymbol{K}, \mathbf{Set})$, is a quotient morphism in $\mathcal{BP}$.

Let us note that there is a similar strengthening of the duality theorem of [M1] for pretoposes.

2'. Theorem. For any small pretopos T, and any $Z : \boldsymbol{K} \longrightarrow \mathbf{Mod} T$ in $\mathbf{UC}$, the 2-category of ultracategories, if Z is full and faithful, then its transpose in the Stone adjunction, $Z^\# : T \longrightarrow \mathcal{H}om(\boldsymbol{K}, \mathbf{Set})$, is a quotient morphism in $\mathcal{PRETOP}$.

The proof of this theorem is essentially contained in [M1]; the minor changes to be made can be seen in the proof given below of Theorem 2. Note that the relation of the duality theorem 4.1 of [M1] to the present 2' is like the relation of the "conceptual completeness" theorem for pretoposes, 7.1.8 in [M/R], to the "strong conceptual completeness" theorem, emphasized by A. Joyal (but in essence, although not explicitly, contained in [M/R] already), which says, in the notation of *loc.cit.*, that $I^* : \mathrm{Mod}(S) \longrightarrow \mathrm{Mod}(P)$ is full and faithful iff $I : P \longrightarrow S$ is a quotient morphism.

Proof of 1. from 2. When, in 2., $\boldsymbol{K} = \mathbf{Mod}^* T$, and $Z = 1_{\mathbf{Mod}^* T}$, then $Z^\# = \varepsilon_T$. Hence, by 2., ε_T is a quotient morphism. By the Gödel/Deligne completeness theorem (see [M/R], or [M3]), ε_T is conservative. By (the straightforward) 2.4.6 in [M3], ε_T is an equivalence of categories. [] 1.

A special case of 2. is obtained when Z is taken to be a full inclusion. In this case, the data for the situation are given, besides T, by a class K of models closed under ultraproducts and ultralimits. This case of 2. is in fact representative; the general case is a consequence of the special case as follows.

With Z as in 2., let L be the replete image of the functor-part of Z (in particular, L is closed under ultraproducts and ultralimits), and let $\boldsymbol{L}$ be the full sub-ultragroupoid of

$\mathbf{Mod}^* T$ with underlying category L, with the E-relations defined in $\boldsymbol{L}$ exactly as in $\mathbf{Mod}^* T$. Then Z is factored as $\boldsymbol{K} \xrightarrow{Z'} \boldsymbol{L} \xrightarrow{Z''} \mathbf{Mod}^* T$, where Z'' is the inclusion. By construction, and the conditions on Z, the functor-part of Z' is an equivalence of categories.

Now, a straightforward version of Lemma 8.4 in [M1] says that for any arrow in $\mathbf{UG}$, if the functor-part of it is an equivalence of categories, then the arrow itself is an equivalence-arrow in the 2-category $\mathbf{UG}$; the proof of this uses the invariance condition imposed on the E-relations in p-u.gr.'s. It follows that Z' is an equivalence in $\mathbf{UG}$, and $F(Z')$ is an equivalence in $\mathcal{BP}$. We have $Z^{\#} = Z'' \circ F(Z')$; the claim follows.

We turn to the proof of 2. in the special case.

By 2.1., it suffices to show that

(*) $Z^{\#}$ is full on subobjects;

and

(**) $Z^{\#}$ is finitely subcovering.

Proof of (*). This is very much like the proof of 4.2(ii) in [M1], on pp. 125-128; we confine ourselves to a sketch; we follow the notation of *loc.cit.* Note that $Z^{\#}: T \longrightarrow \mathcal{H}om(\boldsymbol{K},\mathbf{Set})$ acts as evaluation: $A \longmapsto [M \longmapsto M(A)]$, etc.; in other words, $Z^{\#}(A) = \varepsilon_T(A)$, etc.

Let $v: X \rightarrowtail \varepsilon A$ be a monomorphism in $T' = \mathcal{H}om(\boldsymbol{K},\mathbf{Set})$. Let $X^* M$ be the subset $\mathrm{Im}(v_M)$ of MA $(M \in K)$. Let

$$S = \{\Sigma \in \mathrm{Sub}_T A : X^* N \subset N\Sigma \text{ for all } N \in K\}.$$

Fix $M \in K$. To show

$$(***) \quad \bigcap_{\Sigma \in S} M\Sigma \subset X^* M,$$

let $a \in \bigcap_{\Sigma \in S} M\Sigma$; we want to see that $a \in X^* M$. Let $J = \{\Sigma \in \mathrm{Sub}_T A : a \notin M\Sigma\}$, V an ultrafilter on J with $[\Sigma] \underset{\mathrm{def}}{=} \{\Phi \in J : \Sigma \leq \Phi\} \in J$ for all $\Sigma \in J$. Let $N_\Sigma \in K$, $b_\Sigma \in N_\Sigma A$ such that $b_\Sigma \in X^* N_\Sigma - N_\Sigma \Sigma$; let $N = \prod_{\Sigma \in J} N_\Sigma / V$, $b = \langle b_\Sigma \rangle_{\Sigma \in J} / V$. We have, as in *loc.cit.*, that $b \in N\Sigma$ implies $a \in M\Sigma$ for any $\Sigma \in \mathrm{Sub}_T A$.

Since $\mathrm{Sub}_T A$ is complemented (T is Boolean), we in fact have

$$a \in M\Sigma \iff b \in N\Sigma \quad \text{for all } \Sigma \in \mathrm{Sub}_T A.$$

This means that the structures (M, a) , (N, b) , over the language the underlying graph of T plus a single individual constant **a** of type A , are elementarily equivalent. Therefore, by the Keisler/Kochen isomorphism theorem, there is an ultrafilter sequence $\vec{U}$ such that the ultralimits $(M, a)^{\vec{U}}$, $(N, b)^{\vec{U}}$ are isomorphic. That is to say, there is an isomorphism $h: N^{\vec{U}} \xrightarrow{\cong} M^{\vec{U}}$ such that $\delta_{NA} b = \delta_{MA} a$.

Consider the diagram

$$\begin{array}{ccc}
(\prod N_\Sigma A/V)^{\vec{U}} & \xrightarrow[\cong]{h_A} & MA^{\vec{U}} \\
v_1 \underset{\text{def}}{=} (\prod v_{N_\Sigma}/V)^{\vec{U}} \Big\uparrow & & \Big\uparrow v_2 \underset{\text{def}}{=} (v_M)^{\vec{U}} \\
(\prod XN_\Sigma/V)^{\vec{U}} & \underset{\cong}{\dashrightarrow} & (XM)^{\vec{U}}
\end{array}$$

There is a (unique) function (in fact, bijection) for the dashed arrow so as to make the diagram commute; in other words, h_A restricts to a bijection on the images of the vertical arrows. This is obvious when X is strict (because v is an $\overset{*}{\text{u.t.}}$; now Xh fits for that function); the general case is easily seen by extending the diagram with relevant arrows.

Now a look at the stipulations so far reveals that $b \in \text{Im}(v_1)$. Hence, $\delta_{MA}(a) = h_A(b) \in \text{Im}(v_2)$. Thus there are $n < \omega$, $\vec{U}' \underset{\text{def}}{=} \vec{U} \restriction [1, n]$, $P \in \dot{\times}(\vec{U}')$, and $x = \langle x_{\vec{i}} \rangle_{\vec{i} \in P} / \vec{U}' \in (XM)^{\vec{U}}$ such that

$$(v_M)^{\vec{U}}(x) = \langle v_M(x_{\vec{i}}) \rangle_{\vec{i} \in P} / \vec{U}' = \delta_{MA}(a) = \langle a \rangle_{\vec{i} \in P} / \vec{U}' ;$$

and so, for at least one $\vec{i} \in P$, $v_M(x_{\vec{i}}) = a$, implying that $a \in X^* M$ as desired. This proves (***).

With $M \in K$, by definition, $X^* M \subset M\Sigma$ for all $\Sigma \in S$. Using (***) we can show, as in Claim 2 in *loc.cit.*, that there must be a $\Sigma \in S$ such that $X^* M = M\Sigma$ for all $M \in K$. It is then straightforward to show (as in *loc.cit.*) that the subobject $[X \rightarrowtail z^\# A]$ is the same as $z^\#(\Sigma)$.

[] (*)

The proof of (**) is also similar, in its main outline, to the proof of the corresponding assertion, 4.2.(iii), in [M1]; the latter proof is the main part of the paper [M1]. From now on, $X \in \mathcal{H}om(\mathbf{K},\mathbf{Set})$ is fixed; we are aiming at a finite subcovering of X via $Z^{\#}$.

The notion of support is taken over from [M1] a bit modified. Let $M \in K$, $x \in XM$, $A \in T$ and $a \in MA$. (A, a) is a *support of* x if for any $N \in K$, any ultrafilter sequence $\vec{U}$ of type ω, and any pair of isomorphisms $M^{\vec{U}} \underset{h_2}{\overset{h_1}{\rightrightarrows}} N$, we have that $(h_1)_A(\delta a) = (h_2)_A(\delta a)$ implies that $(Xh_1)(\zeta^{-1}\hat{\delta}x) = (Xh_2)(\zeta^{-1}\hat{\delta}x)$; here we have used the abbreviations $\delta = \delta^{\vec{U}}_{MA}$, $\zeta = [X, \vec{U}]$, $\hat{\delta} = \delta^{\vec{U}}_{XM}$.

3. Lemma. Every $x \in XM$ has some support (A, a) .

Proof. This is the point where the E-relations in the notion of "ultragroupoid" become operative. Otherwise, not unexpectedly, the proof imitates that of the corresponding lemma, 4.4, in [M1].

As in *loc.cit.*, it suffices to have a finite family i of pairs (A, a) such that $h_{1A}\delta a = h_{2A}\delta a$ for all $(A, a) \in i$ implies that $(Xh_1)(\delta' x) = (Xh_2)(\delta' x)$. To arrive at a contradiction, we assume there is no such family i . With $I = \mathcal{P}_{\omega}\{(A, a) : A \in T,\ a \in MA\}$, for every $i \in I$, there are $\vec{U}^i$, $M^{\vec{U}^i} \underset{h_2^i}{\overset{h_1^i}{\rightrightarrows}}{}_{\cong} N_i$ such that $h^i_{1A}\delta^i a = h_{2A}\delta'^{\,i} a$, but $(Xh^i_1)(\delta'^{\,i}x) \neq (Xh^i_2)(\delta'^{\,i}x)$ for all $i \in I$. Let U be an ultrafilter on I such that $\{j \in I : i \subset j\} \in U$ for all $i \in I$, let

$$h_k \underset{\mathrm{def}}{=} \prod_{i \in I} h^i_k/U : M^{\vec{U}} \underset{\mathrm{def}}{=} \prod_{i \in I} M^{\vec{U}^i}/U \xrightarrow{\ \cong\ } \prod_{i \in I} N_i/U$$

for $k = 1, 2$. Let $\delta = \delta^{\vec{U}}_M : M \longrightarrow M^{\vec{U}}$ be the canonical map. We see (as in *loc.cit.*) that $h_1 \circ \delta = h_2 \circ \delta$, *that is,* $(h_1, h_2) \in E^{\vec{U}}_{M,N}$ in the sense of the ultragroupoid $\mathbf{Mod}^* T$. Since X preserves the E-relations, we have $(\xi h_1, \xi h_2) \in E^{\vec{U}}_{XM, XN}$, with

$$\xi : \mathrm{Hom}_{\mathbf{Mod}^* T}(M^{\vec{U}}, N) \longrightarrow \mathrm{Hom}_{\mathbf{Set}^*}((XM)^{\vec{U}}, XN)$$

the mapping induced by X (simply the effect of X in case X is strict). This means that

$$\xi h_1 \circ \delta^{\vec{U}}_{XM} = \xi h_1 \circ \delta^{\vec{U}}_{XM};$$

in particular, the two functions applied to the element $x \in XM$ give the same value. However, the construction ensures exactly that this latter equality fails (the calculation for this is as at the end of the proof in *loc.cit.*). [] 3.

Since we are to show that X is finitely subcovered via $Z^{\#}$, we need to deal with subobjects of appropriate $\mathrm{u}^*\mathrm{f}$.'s. In fact, we could say that the main point of the proof (here of 2., in [M1] of the pretopos duality theorem) is the construction "by hand", or rather, by a transfinite-combinatorial argument, of the necessary subobjects for the subcovering situation. This construction first will produce only small (in the technical sense) parts of the subobjects; in the final step, these parts are fitted together to yield the whole subobjects. We conceptualize those small parts as ultra*diagrams of the form $\mathcal{A}:\Gamma \longrightarrow \mathbf{Set}^*$.

Any subobject Σ of $\mathcal{A}$ in $\mathrm{Hom}(\Gamma, \mathbf{Set})$, $[\rho:\Sigma \rightarrowtail \mathcal{A}]$, is uniquely determined by the family $\langle \Sigma(\gamma) \rangle_{\gamma \in |\Gamma|}$ of the subsets $\Sigma(\gamma) = \mathrm{Im}(\rho_\gamma) \subset \mathcal{A}(\gamma)$. The next lemma uses the concepts introduced in the two previous sections.

4. Lemma. Let $\mathcal{A}:\Gamma \longrightarrow \mathbf{Set}^*$ be an $\mathrm{u}^*\mathrm{d}$., with Γ *small*; let $\gamma_0 \in |\Gamma|$, $x \in \mathcal{A}(\gamma_0)$. Then there is a grounded rooted cell-system $G:\Lambda \longrightarrow \Gamma$ such that $G(\lambda_{\mathrm{in}}) = \gamma_0$, and such that for any filling $\mathcal{F}$ of Λ along $\mathcal{A}$ with $\mathcal{F}(\lambda_{\mathrm{in}}) = x$, the sets

$$\Sigma(\gamma) = \{\mathcal{F}(\lambda) : \lambda \in \mathrm{dom}(\mathcal{F}),\ G(\lambda) = \gamma\} \subset \mathcal{A}(\gamma) \qquad (\gamma \in |\Gamma|)$$

determine a subobject $\Sigma_{\mathcal{F}}$ of $\mathcal{A}$.

Proof. By 6.(v) and 7.(iii). [] 4

Let $\mathcal{M}:\Gamma \longrightarrow \boldsymbol{K}$ be an $\mathrm{u}^*\mathrm{d}$. A *partial A-cover of X relative to $\mathcal{M}$* is a subobject of $(Z^{\#}A \times X) \circ \mathcal{M} : \Gamma \to \mathbf{Set}^*$ in $\mathrm{Hom}(\Gamma, \mathbf{Set})$ such that the composite

$$\Sigma \rightarrowtail (Z^{\#}A \times X) \circ \mathcal{M} \xrightarrow{\pi_1} Z^{\#}A \circ \mathcal{M}$$

is a monomorphism (this means that, for each $\gamma \in |\Gamma|$, $(a, x) \in \Sigma(\gamma)$ and $(a, x') \in \Sigma(\gamma)$ imply $x = x'$).

The following lemma contains the motivation for the notion of ultra$^{(*)}$morphism; in fact, that notion is designed to enable us to carry out the argument that follows.

5. ("main") Lemma. Let $\gamma_0 \in |\Gamma|$, $M = \mathcal{M}(\gamma_0)$, $x \in XM$, and let (A, a) be a support of x. Then there is a partial A-cover Σ of X relative to $\mathcal{M}$ such that $(a, x) \in \Sigma(\gamma_0)$.

Proof. Let $\mathcal{A} = (Z^{\#}A \times X) \circ \mathcal{M}$; let us apply 4., with (a, x) playing the role of x, to get $G:\Lambda \longrightarrow \Gamma$ as there. Assume that the assertion of the lemma fails. Choose and fix a filling $\mathcal{F}$ of Λ along $\mathcal{A}$ with $\mathcal{F}(\lambda_{\mathrm{in}}) = (a, x)$. For any $\Psi \in \mathrm{Reg}(\Lambda)$, consider the filling $\mathcal{F}_\Psi = \mathcal{F}\upharpoonright(\mathrm{dom}(\mathcal{F}) \wedge \Psi)$, and the corresponding subobject $\Sigma_\Psi = \Sigma_{\mathcal{F}_\Psi}$ of $\mathcal{A}$ given in 4. By assumption, Σ_Ψ is not a partial A-cover of X; there are $\gamma_\Psi \in |\Gamma|$, $\lambda_{1\Psi}, \lambda_{2\Psi} \in \Psi$ such that $G(\lambda_{1\Psi}) = G(\lambda_{2\Psi}) = \gamma_\Psi$, and for $(a_{1\Psi}, x_{1\Psi}) = \mathcal{F}(\lambda_{1\Psi})$, $(a_{2\Psi}, x_{2\Psi}) = \mathcal{F}(\lambda_{2\Psi})$, we have $a_{1\Psi} = a_{2\Psi}$, but $x_{1\Psi} \neq x_{2\Psi}$. For $p = 1, 2$, let $\varphi_p:\mathrm{Reg}(\Lambda) \to |\Lambda|$ be the function for which $\varphi_p(\Psi) = \lambda_{p\Psi}$. Let W be any ultrafilter on $\mathrm{Reg}(\Lambda)$ such that $[\Psi] \in W$ for all $\Psi \in \mathrm{Reg}(\Lambda)$. Let S^p be the ultrascheme $(G:\Lambda \longrightarrow \Gamma, \varphi_p, \mathrm{Reg}(\Lambda), W, 1_{\mathrm{Reg}(\Lambda)})$.

S^1 and S^2 are paired ultraschemes; let, by 6.(xx), $\vec{S}^1$ and $\vec{S}^2$ be paired isomorphism ultraschemes extending S^1 and S^2, respectively. Let $v^p = v^{\vec{S}^p}$ be the full $\mathrm{u}^*\mathrm{m}$. on $\mathbf{Set}$ determined by $\vec{S}^p$ according to 7.(vi). Let $\mathcal{M}_\omega:\Gamma_\omega \longrightarrow \boldsymbol{K}$ be the $\mathrm{u}^*\mathrm{d}$. extending $\mathcal{M}$ which is strict for each of the specifications u^+, ℓ_ω, ℓ'_ω. Let us evaluate $v^p_{\boldsymbol{K}}$ at $\mathcal{M}_\omega$, and $v^p = v^p_{\mathbf{Set}}$ at $X\mathcal{M}_\omega:\Gamma_\omega \longrightarrow \mathbf{Set}^*$ on elements that interest us. With $M_\Psi = \mathcal{M}(G\varphi^p(\Psi))$ (no dependence on p) and $N = \prod_{\Psi \in \mathrm{Reg}(\Lambda)} M_\Psi / W$, we get

$$(v^p_{\boldsymbol{K},\mathcal{M}_\omega})_A : M^{\vec{W}}(A) \longrightarrow N^{\vec{W}'}(A)$$

$$\delta(a) \longmapsto \delta'(\langle a_{p\Psi}\rangle_\Psi / W)$$

$$(\delta = (\delta^{\vec{W}}_M)_A,\quad \delta' = (\delta^{\vec{W}'}_N)_A)$$

and

$$v^p_{X\mathcal{M}_\omega} : X(M^{\vec{W}}) \longrightarrow X(N^{\vec{W}'})$$

$$\zeta^{-1}\hat{\delta}(x) \longmapsto \zeta'^{-1}\hat{\delta}'(\langle x_{p\Psi}\rangle_\Psi / W)$$

$$(\hat{\delta} = \delta^{\vec{W}}_{XM},\ \hat{\delta}' = \delta^{\vec{W}'}_{\prod_{\Psi} XM_{\Psi}/W},\ \zeta = [X, \vec{W}]_M,\ \zeta' = [X, W]^{\vec{W}'}_{\langle M_\Psi\rangle_\Psi} \circ [X, \vec{W}']_N).$$

These values are obtained by applying the definition in 7.(vi) first with $\mathcal{A}_\omega = \varepsilon_T A \circ \mathcal{M}_\omega : \Gamma_\omega \to \mathbf{Set}^*$ (a strict $\overset{*}{\text{u}}$.d.), secondly with $\mathcal{A}_\omega = X \circ \mathcal{M}_\omega$; writing $\mathcal{F}(\lambda) = (\mathcal{G}(\lambda), \mathcal{H}(\lambda))$, we have that $\mathcal{G}$, $\mathcal{H}$ are fillings along, respectively, the first and the second $\overset{*}{\text{u}}$.d.; they are the ones to be used.

By the choice of the $a_{p\Psi}$, $x_{p\Psi}$, we get that

$$(v^1_{\mathbf{K},\, \mathcal{M}_\omega})_A(\delta(a)) = (v^2_{\mathbf{K},\, \mathcal{M}_\omega})_A(\delta(a))$$

and

$$v^1_{X\mathcal{M}_\omega}(\zeta^{-1}\hat{\delta}(x)) \neq v^2_{X\mathcal{M}_\omega}(\zeta^{-1}\hat{\delta}(x)) .$$

But since X preserves Δ-$\overset{*}{\text{u}}$.m.'s, $X(v^p_{\mathbf{K},\, \mathcal{M}_\omega}) = v^p_{X\mathcal{M}_\omega}$. With $h_p = v^p_{\mathbf{K},\, \mathcal{M}_\omega}$, we have $(h_1)_A(\delta a) = (h_2)_A(\delta a)$, but $(Xh_1)(\zeta^{-1}\hat{\delta}(x)) \neq (Xh_2)(\zeta^{-1}\hat{\delta}(x))$, in contradiction to (A, a) being a support of x.

[] 5.

The rest of the proof of (**), using 3. and 5., is quite the same as the corresponding part of [M1] (pp. 140-143, including 4.7 and 4.8). In the third line of p. 141, one associates with N a model M with certain properties; in our case, M will have to be chosen in K, and the literal choice in *loc.cit.* may not be possible. However, all what is used in *loc.cit.* is that $M \equiv N$ and that the totality of all the M's chosen is small. In our case, N will be given as a member of K. What we do is look at the equivalence classes of elementary equivalence restricted to K, and fix a choice function selecting a member M_c to each class c. In the passage of the proof in question, we let $M = M_c$ with c the class of N.

We trust that the reader will find the transfer of [M1] to the present situation unproblematic.

[] (**)

[] 2.

9. PREPARING A FUNCTOR SPECIFICATION

This section is a replacement, in the "Boolean" context, for Zawadowski's "representation", or normal form, of descent data; see 5.3 in [Z2]. This latter is a crucial (for Zawadowski), but ultimately simple, fact about arrows in cocomma objects in $\mathcal{P}retop$; it was first noted and used by A. Pitts in [P3]. We will come up with a somewhat more complicated-looking representation of descent data (isomorphism-functor specifications, IFS's, as they are named here); see 3. Proposition below. In particular, whereas Pitts' normal form applies to all arrows in a cocomma object, our normal form is only shown to hold under the additional conditions present in descent data (IFS's).

After reading this section, one might feel that the analysis given here might turn out to be sufficient, essentially by itself, without the rest of the paper, for the proof of the descent theorem, and that some small additional point might be enough to see this. What speaks against this feeling is the circumstance that Pitts' normal form by itself, without Zawadowski's work with ultracategories, does not seem to be enough for the proof of the pretopos descent theorem.

In what follows, our basic data consist of a (many-sorted) theory $\mathcal{T} = (L', T)$ in a language L' , with set of axioms T, and a sublanguage L of L' . To simplify notation, we will use the further assumption that, given any sorts S_i $(i < n)$ in L , there are further sorts $\prod_{i<n} S_i$, $\coprod_{i<n} S_i$ and (unary, sorted) operation symbols $\pi_i : \prod_{i<n} S_i \longrightarrow S_i$, $\iota_i : S_i \longrightarrow \coprod_{i<n} S_i$ $(i < n)$, all in L , such that for all models N of $\mathcal{T}$, $\langle N\pi_i : N(\prod_{i<n} S_i) \longrightarrow NS_i \rangle_{i<n}$ is a product diagram, and $\langle N\iota_i : NS_i \longrightarrow N(\coprod_{i<n} S_i) \rangle_{i<n}$ is a coproduct (disjoint sum) diagram in Set . The additional assumption will hold true when we apply the contents of this section to pretoposes and their internal theories.

$Mod(\mathcal{T})$ is the category of models of $\mathcal{T}$: the category whose objects are the L'-structures that are models of T, and whose arrows are the (L'-)elementary embeddings. " $M \models \mathcal{T}$ " is synonymous with " $M \in Mod(\mathcal{T})$ ". $\mathcal{S}_L$ denotes the set of sorts in L .

Assume the following further data and conditions:

X, a sort in L' ;

for each sort Y in L, $\boldsymbol{f}_Y : Y \longrightarrow Y$ a new unary function symbol (not in L'); let $\boldsymbol{f}$ denote the tuple $\langle \boldsymbol{f}_Y : Y \in \mathcal{S}_L \rangle$;

$L\text{-AUT}[\boldsymbol{f}]$, the set of sentences over $L[\boldsymbol{f}]$ $(= L \cup \{\boldsymbol{f}_Y : Y \in \mathcal{S}_L\})$ expressing that $\boldsymbol{f} = \langle \boldsymbol{f}_Y : Y \in \mathcal{S}_L \rangle$ gives an automorphism of the underlying L-structure: for any L-structure M, $(M, f) \models L\text{-AUT}[\boldsymbol{f}]$ iff $f \in \mathrm{Aut}(M)$ (here $f = \langle f_Y \rangle_{Y \in \mathcal{S}_L}$, and (M, f)

abbreviates the $L[\boldsymbol{f}]$-structure $(M, f_Y)_{Y\in\mathcal{S}_L}$ whenever M is an L-structure, and $f_Y: MY \longrightarrow MY$ interprets $\boldsymbol{f}_Y$ for all $Y\in\mathcal{S}_L$);

$\lambda[\boldsymbol{f}]$, a formula in the language $L'[\boldsymbol{f}]$ (of course, it uses only finitely many of the f_Y) with at most the two distinct variables x , x' , both of sort X, such that

(1) $T\cup L\text{-AUT}[\boldsymbol{f}] \vdash$ " $\lambda[\boldsymbol{f}]$ defines a bijection $X \xrightarrow{\cong} X$ ",

(2) $T \vdash$ " $\lambda[\mathbf{id}]$ defines the identity map on X " ,

(3) $T\cup L\text{-AUT}[\boldsymbol{f}] \cup L\text{-AUT}[\boldsymbol{g}] \vdash$ " $\lambda[\boldsymbol{g}]\circ\lambda[\boldsymbol{f}] = \lambda[\boldsymbol{g}\circ\boldsymbol{f}]$ ".

(Here, the quotes describe informally, but hopefully unambiguously, certain sentences; the first is in the language $L'[\boldsymbol{f}]$, the second in L' , the third in $L[\boldsymbol{f}, \boldsymbol{g}]$ with $\boldsymbol{g} = \langle \boldsymbol{g}_Y : Y \longrightarrow Y \mid Y\in\mathcal{S}_L\rangle$ new operation symbols as indicated. $\lambda[\boldsymbol{g}]$ and $\lambda[\boldsymbol{g}\circ\boldsymbol{f}]$ are obtained from $\lambda[\boldsymbol{f}]$ by replacing each atomic formula $\boldsymbol{f}_Y y = y'$ by $\boldsymbol{g}_Y y = y'$ and by $\boldsymbol{g}_Y(\boldsymbol{f}_Y y) = y'$, respectively .)

Any pair $(X, \lambda[\boldsymbol{f}])$ as described is called a *uniformly definable automorphism-group action* (UDAGA); or more specifically, a $\mathcal{T}/L$-UDAGA.

Given one such, if N is any model of $\mathcal{T}$, $M = N{\restriction}L$, then we have an action of the group $\mathrm{Aut}(M)$ on the set $N(X)$, where the permutation of $N(X)$ corresponding to the automorphism $f = \langle f_Y\rangle_Y$ of M is the interpretation $\lambda[f]$ (or $\lambda_N[f]$) of the formula $\lambda[\boldsymbol{f}]$ in the structure $(N, f_Y)_{Y\in L}$.

1. Lemma. Let us have a $\mathcal{T}/L$-UDAGA $(X, \lambda[\boldsymbol{f}])$. Let $N\models\mathcal{T}$, $M = N{\restriction}L$. Then the action induced on N is *continuous*; that is, for any $x\in N(X)$, there are $n<\omega$, $Y_i\in\mathcal{S}_L$ and $y_i\in MY_i$ for $i<n$ such that for any $f\in\mathrm{Aut}(M)$, if $f_{Y_i} y_i = y_i$ for $i<n$, then $\lambda[f](x) = x$.

Proof. For simplicity, and without loss of generality, we assume that the sets MY ($Y\in\mathcal{S}_L$) are pairwise disjoint; we put $A \underset{\text{def}}{=} \bigcup_{Y\in\mathcal{S}_L} MY$. Each $f\in\mathrm{Aut}(M)$ can be construed as a permutation of A . Let P be the set of all finite sets $p\subset A\times A$ such that there is $f\in\mathrm{Aut}(M)$ with $p\subset\mathrm{graph}(f)$. Let us partially order P by $q\le p \underset{\text{def}}{\Longleftrightarrow} p\subset q$.
Let $\boldsymbol{B}$ be the complete Boolean algebra of regular subsets of $(P, \le)$; $B\subset P$ is *regular* if, for any $p\in P$, $p\in B$ iff $\forall q{\le}p\,\exists r{\le}q\; r\in B$; regular sets are in particular closed downward. Let $\mathcal{E}$ be the category of $\boldsymbol{B}$-valued sets (see e.g. Chapter 4 of [M/R]). $\mathcal{E}$ is a Boolean topos; in particular, it is a Boolean pretopos.

Each object Z of $\mathcal{E}$ is a pair $Z = (\,|Z|\,,\ [\![=_Z]\!]\,)$ with $[\![=_Z]\!] : |Z|\times|Z| \longrightarrow \boldsymbol{B}$; we write $p\Vdash z = z'$ for $p\in [\![z =_Z z']\!]$. A *$\boldsymbol{B}$-valued predicate on* Z is a function

$\Phi: |Z| \longrightarrow \mathcal{B}$ satisfying

$$[\![z=_Z z']\!] \wedge \Phi(z) \leq \Phi(z')$$

and

$$\Phi(z) \leq [\![z=_Z z]\!]$$

for all $z, z' \in |Z|$.

If $\hat{L}$ is any (first order) language, $\hat{M}: \hat{L} \longrightarrow \mathcal{E}$ an interpretation of L in $\mathcal{E}$, $\varphi(\vec{x})$ an L-formula with the indicated free variables, $\vec{a}$ a tuple "matching" $\vec{x}$ (for $\vec{x} = \langle x_i \rangle_{i<n}$, x_i of sort X_i, we have that $\vec{a} = \langle a_i \rangle_{i<n}$, and $a_i \in |\hat{MX}_i|$), $[\![\varphi[\vec{a}]]\!]_{\hat{M}}$, or just $[\![\varphi[\vec{a}]]\!]$, denotes the (*Boolean*) *value* of φ at $\vec{a}$ in $\hat{M}$. (The subobject of $\prod_{i<n} \hat{M}X_i$ which is the interpretation of φ in $\hat{M}$ is given by the $\mathcal{B}$-valued predicate $\vec{a} \longmapsto [\![\varphi[\vec{a}]]\!]$.) We write $p \Vdash \varphi[\vec{a}]$ for $p \in [\![\varphi[\vec{a}]]\!]$.

If Σ is a set of $\hat{L}$-sentences, $\hat{M}: L \longrightarrow \mathcal{E}$, then $\hat{M} \models \Sigma$ abbreviates the fact that $[\![\tau]\!]_{\hat{M}} = 1_{\mathcal{B}}$ for all $\tau \in \Sigma$. If $\Sigma \vdash \sigma$, and $\hat{M} \models \Sigma$, then $\hat{M} \models \sigma$, as a consequence of the fact that $\mathcal{E}$ is a Boolean pretopos.

A $\mathcal{B}$-set Z is *discrete* if $[\![z=_Z z']\!] = 1_{\mathcal{B}}$ when $z = z'$, $[\![z=_Z z']\!] = 0_{\mathcal{B}}$ otherwise; and a $\mathcal{B}$-valued predicate on a discrete $\mathcal{B}$-set is *discrete* if it takes only the values $0_{\mathcal{B}}$ and $1_{\mathcal{B}}$. Any ordinary **Set**-valued interpretation $\hat{M}$ of L can be made, in the obvious way, into a $\mathcal{B}$-valued interpretation, also denoted by $\hat{M}$ and called itself *discrete*, all whose sorts and predicates are discrete. In this case, we will have $\hat{M} \models \varphi[\vec{a}]$ (in the ordinary sense) iff $[\![\varphi[\vec{a}]]\!]_{\hat{M}} = 1_{\mathcal{B}}$.

Any automorphism ρ of the poset $(P, \leq)$ induces an automorphism of $\mathcal{E}$: for a $\mathcal{B}$-set Z, $\rho' Z$ is the $\mathcal{B}$-set for which $|\rho' Z| = |Z|$, and $p \Vdash z =_{\rho' Z} z' \iff \rho^{-1}(p) \Vdash z =_Z z'$; and for any $\mathcal{B}$-valued predicate Φ on Z, $p \Vdash (\rho'\Phi)(z) \iff \rho^{-1}(p) \Vdash \Phi(z)$ (the action of ρ on morphisms is given by its action on the graphs of the morphisms as predicates). Note that on discrete $\mathcal{B}$-sets and $\mathcal{B}$-valued predicates, ρ acts as the identity. If $\hat{M}: L \longrightarrow \mathcal{E}$, then we have

$$\rho' \hat{M} \underset{\text{def}}{=} \hat{M} \circ \rho : L \longrightarrow \mathcal{E};$$

and obviously (or by induction on the structure of the formula φ if any doubt arises),

$$p \Vdash_{\hat{M}} \varphi[\vec{a}] \iff \rho(p) \Vdash_{\rho'\hat{M}} \varphi[\vec{a}] .$$

Now, let us return to our specific P, $\mathcal{B}$ and $\mathcal{E}$. Let $f \in \mathrm{Aut}(M)$. If $p \in P$, let $p \circ f$ denote $\{(a, c) : (fa, c) \in p\}$; clearly, $p \circ f \in P$, in particular, $p \circ f \subset g \circ f$ if $p \subset g \in \mathrm{Aut}(M)$. It is clear too that $p \longmapsto p \circ f$ is an automorphism ρ of $(P, \leq)$; we write $f'(-)$ for $\rho'(-)$ in the contexts above when $\rho'(-)$ was used.

Now, let us concentrate on the given structures N and M. They give rise to discrete $\mathcal{B}$-valued structures N and M, resp. The interpretations of L in $\mathcal{E}$ form a category; the group of automorphisms in this category of M is denoted $\mathrm{Aut}_{\mathcal{B}}(M)$ (" $\mathcal{B}$-valued automorphisms"). In particular, each $f \in \mathrm{Aut}(M)$ gives rise, in the obvious way, to a $\mathcal{B}$-valued automorphism of M; in fact, $\mathrm{Aut}(M)$ is a subgroup of $\mathrm{Aut}_{\mathcal{B}}(M)$.

There is a specific $\mathcal{B}$-valued automorphism G of M, given by a predicate $[\![Ga = b]\!]$ on the discrete $\mathcal{B}$-set $A \times A$; the latter predicate is defined by

$$p \Vdash Ga = b \underset{\text{def}}{\iff} pa \simeq b \ (\underset{\text{def}}{\iff} (a, b) \in p).$$

Let us check, for instance, that $0 \Vdash \forall a{\in}A \exists b{\in}A\ Ga = b$. This requires showing that for all $a \in A$, we have $\forall p{\in}P \exists q{\leq}p\ \exists b{\in}A\ q \Vdash Ga = b$. And indeed, if $p \in P$, let $f \in \mathrm{Aut}(M)$ with $p \subset f$, let $b = f(a)$, and let $q = p \cup \{(a, b)\}$; then $q \leq p$ and $q \Vdash Ga = b$. The rest of the verification of the fact that $G \in \mathrm{Aut}_{\mathcal{B}}(M)$ is left to the reader.

Let $f \in \mathrm{Aut}(M)$. As we noted, f acts as an automorphism of the whole category $\mathcal{E}$, and in particular, $f'M = M$, and $f'G \in \mathrm{Aut}_{\mathcal{B}}(M)$. On the other hand, f is (gives rise to) an element of $\mathrm{Aut}_{\mathcal{B}}(M)$, and as such, gives sense to the composite $G \circ f^{-1}$ understood in $\mathrm{Aut}_{\mathcal{B}}(M)$. We claim that in fact $f'G = G \circ f^{-1}$. Indeed,

$$\begin{aligned}
p \Vdash (f'G)a = c &\iff p \circ f^{-1} \Vdash Ga = c && \text{(definition of } f'G\text{)}\\
&\iff (p \circ f^{-1})a \simeq c && \text{(definition of } G\text{)}\\
&\iff p(f^{-1}(a)) \simeq c &&\\
&\iff p \Vdash G(f^{-1}(a)) = c && \text{(definition of } G\text{)}\\
&\iff p \Vdash (G \circ f^{-1})a = c && \text{(as it is easily seen).}
\end{aligned}$$

Now, let us fix $x \in N(X)$. Let $\lambda[G]$ denote the interpretation of $\lambda[\boldsymbol{f}]$ in the $\mathcal{B}$-valued structure (N, G); similarly for other uses of $\lambda[-]$; $\lambda[G]$ is a $\mathcal{B}$-valued predicate on the discrete set $N(X)$, and by (1) above, since $(N, G) \models T \cup L\text{-AUT}[\boldsymbol{f}]$, we have that

$$0 \Vdash \text{" } \lambda[G] \text{ is a bijection } X \xrightarrow{\cong} X \text{".} \tag{4}$$

In particular, there is $p \in P$ and $x' \in N(X)$ such that

$$p \Vdash (\lambda[G])x = x' . \tag{5}$$

Let $I = \mathrm{dom}(p)$ and assume that f is any automorphism of M such that $f(a) = a$ for all $a \in I$. Then $p \circ f \Vdash (\lambda[f'G])x = x'$, that is,

$$p \Vdash (\lambda[G \circ f^{-1}])x = x' ,$$

hence, by (1), (2) and (3),

$$p \Vdash (\lambda[G] \circ (\lambda[f])^{-1})x = x' . \tag{6}$$

Of course, $\lambda[f]^{-1}$ makes sense as the ordinary permutation of the set $N(X)$, obtained by interpreting λ in (N, f), and we get the value $(\lambda[f]^{-1})x \in N(X)$. The permutation $\lambda[f]^{-1}$ gives rise to the discrete $\mathcal{B}$-valued permutation of $N(X)$, and $\lambda[f]^{-1}$ in (6) occurs with the latter meaning. Therefore, (6) is equivalent to

$$p \Vdash (\lambda[G])((\lambda[f]^{-1})x) = x' .$$

Using (5), and (4) again, we conclude

$$p \Vdash (\lambda[f]^{-1})x = x .$$

Since we are entirely in the realm of the discrete here,

$$(\lambda[f]^{-1})x = x ,$$

i.e.,

$$(\lambda[f])x = x . \tag{7}$$

We conclude that (7) holds for all $f \in \mathrm{Aut}(M)$ for which f leaves fixed each element of the specific finite set I, which shows the assertion of the lemma. [] 1.

The theory $\mathcal{T} +_L \mathcal{T}$ is, by definition, the theory (in full first order logic) whose typical model is made up of two models of $\mathcal{T}$ and an isomorphism between their L-reducts. Its language L'' has two copies of each symbol in L'; the sorts of L'' are $X_{(0)}$ and $X_{(1)}$ for sorts X of L', and the relation symbols are $R_{(j)} \subset \prod_{i<n} X_{i(j)}$, one for each $j \in \{0, 1\}$ and $R \subset \prod_{i<n} X_i$ in L'; similarly for operation symbols; in addition, L'' has the operation symbols $\boldsymbol{f}_Y : Y_{(0)} \longrightarrow Y_{(1)}$, one for each $Y \in \mathcal{S}_L$. The axioms of $\mathcal{T} +_L \mathcal{T}$ are the (0)- and the (1)-copies of the axioms of $\mathcal{T}$, and the axioms that assert that, in any model P of them, " $\boldsymbol{f} = \langle \boldsymbol{f}_Y \rangle_Y$ is a isomorphism $P_{(0)} \xrightarrow{\cong} P_{(1)}$ ", with $P_{(0)}$, $P_{(1)}$ the $L_{(0)}$- and the $L_{(1)}$-reduct, respectively, of P understood as L'-structures.

We have the similarly built theory $\mathcal{T}+_L\mathcal{T}+_L\mathcal{T}$, in a language L''', the theory of three models $Q_{(0)}$, $Q_{(1)}$ and $Q_{(2)}$ of $\mathcal{T}$, with isomorphisms $\mathbf{f}_{01}=\langle \mathbf{f}_{01Y}\rangle_Y : Q_{(0)}\restriction L \xrightarrow{\cong} Q_{(1)}\restriction L$, $\mathbf{f}_{12}=\langle \mathbf{f}_{12Y}\rangle_Y : Q_{(1)}\restriction L \xrightarrow{\cong} Q_{(2)}\restriction L$. For a formula $\lambda[\mathbf{f}]$ of $\mathcal{T}+_L\mathcal{T}$, $\lambda[\mathbf{f}_{01}]$ denotes the formula of $\mathcal{T}+_L\mathcal{T}+_L\mathcal{T}$ obtained by replacing $\mathbf{f}$ by $\mathbf{f}_{01}$; $\lambda[\mathbf{f}_{12}]$ is obtained similarly, but now we also replace each $S_{(0)}$ by $S_{(1)}$, and each $S_{(1)}$ by $S_{(2)}$, for each $S\in L'$; there is a corresponding replacement of both bound and free variables to match sorts.

On the other hand, $\bar{\lambda}[\mathbf{f}]$ is the $L'[\mathbf{f}]$-formula obtained by replacing both $S_{(0)}$ and $S_{(1)}$ ($S\in L'$) by S itself, and making each $\mathbf{f}_Y$ a symbol of the type $\mathbf{f}_Y : Y\longrightarrow Y$. $\bar{\lambda}[\mathbf{id}]$ is the L'-formula obtained from $\bar{\lambda}[\mathbf{f}]$ as in (2) above.

An *isomorphism-functor specification for* $\mathcal{T}/L$ ($\mathcal{T}/L$-IFS) is given, along with $\mathcal{T}$ and L, by an $X\in\mathcal{S}_{L'}$ and a formula $\lambda[\mathbf{f}]$ of $\mathcal{T}+_L\mathcal{T}$ with the two distinct free variables $\mathbf{x}$, $\mathbf{x}'$ of sorts $X_{(0)}$ and $X_{(1)}$, resp., such that

(8) $\mathcal{T}+_L\mathcal{T}\vdash$ " $\lambda[\mathbf{f}]$ defines a bijection $X_{(0)}\xrightarrow{\cong}X_{(1)}$ ",

(9) $\mathcal{T}\vdash$ " $\bar{\lambda}[\mathbf{id}]$ defines the identity $X\xrightarrow{\cong}X$ ",

(10) $\mathcal{T}+_L\mathcal{T}+_L\mathcal{T}\vdash$ " $\lambda[\mathbf{f}_{12}]\circ\lambda[\mathbf{f}_{01}]=\lambda[\mathbf{f}_{12}\circ\mathbf{f}_{01}]$ ".

Let $\mathrm{Mod}^*(\mathcal{T}/L)$ denote the category (groupoid) whose objects are the models of $\mathcal{T}$, and whose arrows $M\stackrel{\cdot}{\longrightarrow}N$ are the isomorphisms $M\restriction L\xrightarrow{\cong}N\restriction L$, with composition defined in the obvious way. Note that any $\mathcal{T}/L$-IFS gives rise to a functor

$$[X,\lambda[\mathbf{f}]] : \mathrm{Mod}^*(\mathcal{T}/L)\longrightarrow \mathbf{Set}\ ; \tag{11}$$

the effect of the functor on objects is $M\longmapsto M(X)$, on arrows

$$(h:M\restriction L\xrightarrow{\cong}N\restriction L):M\stackrel{\cong}{\stackrel{\cdot}{\longrightarrow}}N\longmapsto \lambda_{(M,N)}[h] : M(X)\xrightarrow{\cong}N(X)\ ;$$

of course, $\lambda_{(M,N)}[h]$ is the interpretation of $\lambda[\mathbf{f}]$ in $(M,N,h:M|L\xrightarrow{\cong}N|L)$, the latter understood as a structure for L''. Any functor of the form (11) is said to be a *definable functor*.

Note also that an IFS is something more special than an UDAGA. Given the $\mathcal{T}/L$-IFS as above, $\bar{\lambda}[\mathbf{f}]$ gives an UDAGA; however, not every UDAGA is obtained in this way, since in $\bar{\lambda}$ there is a stratification of the variables, a remnant of the "2-sortedness" of λ. Specifically, let us note an invariance property of an UDAGA obtained form an ISF. We say that the $\mathcal{T}$/L-UDAGA $(X,\lambda[f])$ is *invariant* if the following holds: whenever $N\models\mathcal{T}$, $g,h\in\mathrm{Aut}(N)$, $f\in\mathrm{Aut}(N\restriction L)$ and $x,x'\in N(X)$, we have

$$\lambda[f](x) = x' \iff \lambda[(h\upharpoonright L)\circ f\circ(g^{-1}\upharpoonright L)](gx) = hx' .$$

The fact that an UDAGA obtained from an IFS is invariant is just to say that the meaning of formulas in L'' is invariant under isomorphisms of L''-structures.

For *special* models, see [C/K]. I will use the concept of *λ-special model* in the specific way stated in 1.1 of [M6], and the remarks following. Also, an arbitrary "special cardinal" λ is fixed, and "special" will mean "λ-special".

2. Proposition. Let (X, λ) be an invariant $\mathcal{T}/L$-UDAGA, and let N be a special L'-model. Then there are $Y \in \mathcal{S}_L$ and L'-formulas $r(\boldsymbol{y}, \boldsymbol{x})$ and $s(\boldsymbol{x}, \boldsymbol{y})$ $(\boldsymbol{x}:X,\ \boldsymbol{y}:Y)$ such that for all $f \in \mathrm{Aut}(N\upharpoonright L)$,

$$(N, f) \models \forall\boldsymbol{x}\forall\boldsymbol{x}'(\lambda[\boldsymbol{f}](\boldsymbol{x}, \boldsymbol{x}') \longleftrightarrow \exists\boldsymbol{y}(s(\boldsymbol{x}, \boldsymbol{y}) \wedge r(f_Y(\boldsymbol{y}), \boldsymbol{x}'))) .$$

Proof. Consider an arbitrary element $x \in NX$. Let, by 1., (Y, y) be a *support for* x, that is, $Y \in \mathcal{S}_L$, $y \in MY$, and $fy = y \Longrightarrow (\lambda[f])x = x$ for all $f \in \mathrm{Aut}(N\upharpoonright L)$ (let $Y = \prod_{i<n} Y_i$, $y = \langle y_i\rangle_{i<n}$, with the data given by 1.). Let $S(\boldsymbol{x}, \boldsymbol{y})$ be the type of the pair (x, y) in the model N, that is, the infinite conjunction of all L'-formulas $s(\boldsymbol{x}, \boldsymbol{y})$ true of (x, y) in N. Let $f \in \mathrm{Aut}(N\upharpoonright L)$ and $x' = \lambda[f](x)$, $y' = f(y)$, and let $R(\boldsymbol{y}', \boldsymbol{x}')$ be the N-type of (y', x'). Let us consider the following infinitary formula

$$\Phi[\boldsymbol{f}](\boldsymbol{x}, \boldsymbol{x}') \equiv \exists\boldsymbol{y}(S(\boldsymbol{x}, \boldsymbol{y}) \wedge R(\boldsymbol{f}(\boldsymbol{y}), \boldsymbol{x}')) . \tag{12}$$

This formula, according to its construction, depends on the data (x, f); we remove the dependence on y, by picking a fixed support $(Y, y) = (Y_x, y_x)$ for each $x \in N(X)$, and using it in the definition of Φ. Thus, we write $\Phi_{(x, f)}[\boldsymbol{f}]$ for $\Phi[\boldsymbol{f}]$. We claim that for all $f \in \mathrm{Aut}(N\upharpoonright L)$, we have

$$(N, f) \models \forall\boldsymbol{x}\forall\boldsymbol{x}'(\lambda[\boldsymbol{f}](\boldsymbol{x}, \boldsymbol{x}') \longleftrightarrow \bigvee\{\Phi_{(x, f)}[\boldsymbol{f}](\boldsymbol{x}, \boldsymbol{x}') : x \in NX,\ f \in \mathrm{Aut}(N\upharpoonright L)\} . \tag{13}$$

In fact, the assertion with the formula after $\models$ modified by replacing $\longleftrightarrow$ with $\longrightarrow$ is clear from the construction of the $\Phi_{(x, f)}$. Conversely, let x, y, f, x', y' be as above, $\Phi = \Phi_{(x, f)}$, assume that $\hat{x}, \hat{x}' \in NX$, $\hat{f} \in \mathrm{Aut}(N\upharpoonright L)$, $(N, \hat{f}) \models \Phi[\boldsymbol{f}](\hat{x}, \hat{x}')$; we want to show that $\lambda[\hat{f}](\hat{x}) = \hat{x}'$. By assumption, we have $\hat{y} \in NY$ such that

$$S(\hat{x}, \hat{y}) \wedge R(\hat{f}(\hat{y}), \hat{x}')$$

holds. Since N is special, and (x, y) and $(\hat{x}, \hat{y})$ have the same type in N, there is $g \in \mathrm{Aut}(N)$ such that

$$gx = \hat{x}, \quad gy = \hat{y}; \tag{14}$$

similarly, there is $h \in \mathrm{Aut}(N)$ such that

$$hy' = \hat{y}', \quad hx' = \hat{x}'. \tag{15}$$

Let $f' \underset{\mathrm{def}}{=} h^{-1}\hat{f}g$, that is

$$\hat{f} = hf'g^{-1}. \tag{16}$$

Immediately, $f'y = y' = fy$; since y is a support for x, $\lambda[f']x = \lambda[f]x = x'$. By the invariance of the UDAGA (X, λ), and (14), (15), (16), it follows that $\lambda[\hat{f}]\hat{x} = \hat{x}'$, what was to be shown.

A compactness argument will now finish the proof. Let W be the set of all pairs (x, f) with $x \in NX$ and $f \in \mathrm{Aut}(N|L)$. For $w = (x, f) \in W$, let us write S_w, R_w for the formulas S, R, resp., in (12) with $\Phi = \Phi_w$. A *choice function*, for the present purposes, is any function γ with domain W such that for each $w \in W$, $\gamma w = (s_w, r_w) = (s^{\gamma}_w, r^{\gamma}_w)$ with s_w a finite sub-conjunction of the conjunction S_w, and similarly for r_w with respect to R_w. We claim that for any choice function γ,

$$\mathrm{Th}_{L'}(N) \cup L\text{-AUT}[\boldsymbol{f}] \cup \lambda[\boldsymbol{f}](\boldsymbol{x}, \boldsymbol{x}') \models \bigvee_{w \in W} \exists \boldsymbol{y}(s^{\gamma}_w(\boldsymbol{x}, \boldsymbol{y}) \wedge r^{\gamma}_w(\boldsymbol{f}(\boldsymbol{y}), \boldsymbol{x}')); \tag{17}$$

here we use $\boldsymbol{x}$ and $\boldsymbol{x}'$ as new individual constants of type X.

Indeed, the formula after $\models$ is true in (N, f, x, x') for the specific N we have and for any f, x, x' such that (N, f, x, x') is a model of the theory Σ before $\models$; this is because of (13). But then, if $\hat{M} = (M, \hat{f}, \hat{x}, \hat{x}')$ is any model of Σ, there is a special model elementarily equivalent to $\hat{M}$; by the uniqueness of special models, the latter may be chosen to be of the form (N, f, x, x') with the given N. In this, the formula is true; by the elementary equivalence, it is true in $\hat{M}$ as well. This shows the claim.

By the compactness theorem, there is a finite sub-disjunction of the disjunction in (17) which also follows from Σ. We conclude that there is W_{γ}, a finite subset of W such that

$$\Sigma \models \bigvee_{w \in W_{\gamma}} \exists \boldsymbol{y}(s^{\gamma}_w(\boldsymbol{x}, \boldsymbol{y}) \wedge r^{\gamma}_w(\boldsymbol{f}(\boldsymbol{y}), \boldsymbol{x}')). \tag{18}$$

Let us fix a choice of the set W_γ for each γ, and let us call the last formula in (18) $\lambda_\gamma[\boldsymbol{f}](\boldsymbol{x}, \boldsymbol{x}')$. I claim that

$$\mathrm{Th}_{L'}(N) \cup L\text{-AUT}[\boldsymbol{f}] \cup \{\lambda_\gamma[\boldsymbol{f}](\boldsymbol{x},\boldsymbol{x}') : \gamma \text{ a choice function}\} \qquad (19)$$
$$\models \lambda[\boldsymbol{f}](\boldsymbol{x}, \boldsymbol{x}') .$$

Indeed, as before, it suffices to show that for any $\hat{x}$, $\hat{x}' \in NX$ and $\hat{f} \in \mathrm{Aut}(N\restriction L)$ such that $(N, \hat{f})$ is special, if all $\lambda_\gamma[\hat{f}](\hat{x}, \hat{x}')$ are true, then $\lambda[\hat{f}](\hat{x}, \hat{x}')$ is true as well. Assume the last assertion fails at some $\hat{x}$, $\hat{x}'$ and $\hat{f}$. Then, by (13), for any $w \in W$,

$$\Phi_w[\hat{f}](\hat{x}, \hat{x}') \equiv \exists \boldsymbol{y}(S_w(\hat{x}, \boldsymbol{y}) \wedge R_w(\hat{f}(\boldsymbol{y}), \hat{x}'))$$

fails. However, by the fact that (N, f) is $\aleph_0$-saturated, this means that, for any $w \in W$, there are finite subconjunctions s_w, r_w of S_w, R_w, resp., such that $\exists \boldsymbol{y}(s_w(\hat{x}, \boldsymbol{y}) \wedge r_w(\hat{f}(\boldsymbol{y}), \hat{x}'))$ fails as well; and this holds for all $w \in W$. Using the choice function γ for which $w \longmapsto (s_w, r_w)$, we get that, for this γ, $\lambda_\gamma[\hat{f}](\hat{x}, \hat{x}')$ fails (in fact, the weaker infinite disjunction over the whole of W fails as well). This is a contradiction. (19) is thus proved.

Again by compactness, there is a finite set Γ of γ's such that, in (19), on the left side only the γ's in Γ are needed. (18) and (19) thus strengthened together say that

$$\mathrm{Th}_{L'}(N) \cup L\text{-AUT}[\boldsymbol{f}] \models \lambda[\boldsymbol{f}](\boldsymbol{x}, \boldsymbol{x}') \longleftrightarrow \bigwedge_{\gamma\in\Gamma} \lambda_\gamma[\boldsymbol{f}](\boldsymbol{x}, \boldsymbol{x}') . \qquad (20)$$

Let us temporarily call a formula *normal* if it is of the form $\exists \boldsymbol{y}(s(\boldsymbol{x}, \boldsymbol{y}) \wedge r(\boldsymbol{f}(\boldsymbol{y}), \boldsymbol{x}')$, with the fixed free variables $\boldsymbol{x}:X$, $\boldsymbol{x}':X$, with arbitrary $Y \in \mathcal{S}_L$, $\boldsymbol{y}:Y$, and arbitrary L'-formulas $s(\boldsymbol{x}, \boldsymbol{y})$, $r(\boldsymbol{y}, \boldsymbol{x})$. λ_γ in (20) is a disjunction of normal formulas. To finish the proof of the proposition, it suffices to show that any finite disjunction, and any finite conjunction, of normal formulas is equivalent to a normal formula in the theory $T \cup L\text{-AUT}[\boldsymbol{f}]$.

Consider $\sigma(x, x') \equiv \bigvee_i \exists y_i(s_i(x, y) \wedge r_i(f_{Y_i}(y), x'))$. Let $Y = \coprod_i Y_i \in L$, and $\iota_i : Y_i \longrightarrow Y$ the canonical injections. Let, with $y:Y$,

$$s(x, y) \equiv \bigvee_i \exists y_i(s_i(x, y_i) \wedge \iota_i y_i = y) ,$$

$$r(y, x) \equiv \bigvee_i \exists y_i(r_i(y_i, x) \wedge \iota_i y_i = y) .$$

Then $T \cup L\text{-AUT}[\boldsymbol{f}] \models \exists y(s(x, y) \wedge r(f_Y(y), x))$; the verification is left to the reader.

Let $\sigma(x, x') \equiv \bigwedge_i \exists y_i(s_i(x, y) \wedge r_i(f_{Y_i}(y), x'))$. Let $Y = \prod_i Y_i$, $\pi_i : Y \longrightarrow Y_i$ the projections. Let, with $y: Y$,

$$s(x, y) \equiv \bigwedge_i s_i(x, \pi_i y) ,$$

$$s(y, x) \equiv \bigwedge_i r_i(\pi_i y, x) ,$$

and we have a similar conclusion as before. [] 2

3. Proposition. Let $(X, \lambda[\boldsymbol{f}])$ be a $\mathcal{T}/L$-IFS. Let $p(\boldsymbol{x})$ be a complete 1-type in $\mathcal{T}$ (a complete extension of $\mathcal{T}$ with one additional individual constant $\boldsymbol{x}: X$). Then there are $Y \in L$, $\boldsymbol{y}: Y$, an L'-formula $s(\boldsymbol{x}, \boldsymbol{y})$ such that $\exists \boldsymbol{y} s(\boldsymbol{x}, \boldsymbol{y}) \in p$ and the following holds. For any complete 2-type $q(\boldsymbol{x}, \boldsymbol{y})$ with $s \in q$ and $p \subset q$, there is an L'-formula $r(\boldsymbol{y}, \boldsymbol{x})$ such that:

(*) for any $M, N \models \mathcal{T}$, $f: M|L \xrightarrow{\cong} N|L$, $x \in M(X)$, $y \in M(Y)$ such that $M \models q[x, y]$, we have that for any $x' \in N(X)$,

$$\lambda[f](x) = x' \iff N \models r[f_Y y, x'] .$$

Proof. Start with a special model $M \models \mathcal{T}$ in which p is realized; let $x \in MX$ realize p. Consider the invariant $\mathcal{T}/L$-UDAGA $(X, \bar{\lambda}[\boldsymbol{f}])$ induced by (X, λ) (see before Proposition 2). Use Proposition 2, and let $s_1(\boldsymbol{x}, \boldsymbol{y})$, $r_1(\boldsymbol{y}, \boldsymbol{x})$ be the s and r given by that proposition. Let $s(\boldsymbol{x}, \boldsymbol{y}) \equiv s_1(\boldsymbol{x}, \boldsymbol{y}) \wedge r_1(\boldsymbol{y}, \boldsymbol{x})$. To see that $\exists \boldsymbol{y} s[x, \boldsymbol{y})$ holds, consider $\bar{\lambda}[1_M]$; by assumption, $(\bar{\lambda}[1_M])x = x$, and by the choice of s_1 and r_1, $M \models \exists \boldsymbol{y}(s_1(x, \boldsymbol{y}) \wedge r_1(1_Y \boldsymbol{y}, x))$, as desired.

Let us note the following property of s relative to p:

(**) whenever $\hat{M}, N \models \mathcal{T}$, $\hat{x}$ realizes p in $\hat{M}$, $y \in \hat{M}Y$ and

$$\hat{M} \models s[\hat{x}, y] , \quad \hat{M}|L \underset{g}{\overset{f}{\underset{\longrightarrow}{\overset{\longrightarrow}{\cong}}}} N|L \text{, then}$$

$$f_Y y = g_Y y \Rightarrow (\lambda_{\hat{M}, N}[f])\hat{x} = (\lambda_{\hat{M}, N}[g])\hat{x} .$$

First, take $\hat{M} = M$, and $\hat{x} = x$. Let $y \in MY$, $M \models s[x, y]$, $M|L \underset{g}{\overset{f}{\underset{\longrightarrow}{\overset{\longrightarrow}{\cong}}}} N|L$, $f_Y y = g_Y y$. Consider the automorphism $h = g^{-1} \circ f$ of $M \restriction L$, and note that $h_Y y = y$. Let us calculate $x' = (\bar{\lambda}_M[h])x$. By the choice of s, r (see Proposition 2), x' is determined by the property $\exists\boldsymbol{y}(s_1(x,\boldsymbol{y})\wedge r_1(\boldsymbol{y},x'))$. However, this is true for $x'=x$, with y witnessing the variable $\boldsymbol{y}$. Thus $x'=x$. We conclude that $x=(\lambda[g^{-1}\circ f])x=(\lambda[g]^{-1})(\lambda[f])x$, which means $\lambda[g]x=\lambda[f]x$ as desired.

The general case follows by an easy consideration. Note that the property in question is part of the elementary type of the structure $P=(\hat{M},N,f,g,x,y)$. Given any such, with the assumptions for the property holding, clearly P is elementarily equivalent to another such structure in which the $\hat{M}$-component is equal to M; this takes care of the general case.

Now, let $q(\boldsymbol{x},\boldsymbol{y})$ be any complete 2-type in $\mathcal{T}$ such that $p\cup\{s(\boldsymbol{x},\boldsymbol{y})\}\subset q$. Consider the following second order property $\mathcal{P}$ of a structure of the form (N,y',x') of the type $L'(\boldsymbol{y}',\boldsymbol{x}')$, $\boldsymbol{y}'$ a constant of type Y, $\boldsymbol{x}'$ a constant of type X:

$\mathcal{P}$:: " $N\models\mathcal{T}$, and there exist $M\models\mathcal{T}$, $x\in MX$, $y\in MY$ and $f:M|L\xrightarrow{\cong}N|L$ such that $M\models q[x,y]$, $f_Y y=y'$ and $\lambda[f]x=x'$."

Let $R(\boldsymbol{y}',\boldsymbol{x}')$ be the (infinite) conjunction of the set of all $L'(\boldsymbol{y}',\boldsymbol{x}')$-sentences $\rho(\boldsymbol{y}',\boldsymbol{x}')$ which are consequences of the property $\mathcal{P}$: whenever $(N,y',x')\models\mathcal{P}$, then $(N,y',x')\models\rho$. When in (*), we replace r by R, and $\Longleftrightarrow$ by $\Longrightarrow$, the resulting condition $(*)_{\Longrightarrow,R}$ is obviously true.

We claim that

$$R(\boldsymbol{y}',\boldsymbol{x}')\wedge R(\boldsymbol{y}',\boldsymbol{x}'')\models\boldsymbol{x}'=\boldsymbol{x}''. \tag{21}$$

It suffices to show this for an arbitrary special model $P=(N,y',x',x'')$. Assume $P\models R(\boldsymbol{y}',\boldsymbol{x}')\wedge R(\boldsymbol{y}',\boldsymbol{x}'')$. By the definition of R, and the relation universality of special models, there are, for $i=1,2$, $M_i\models\mathcal{T}$, $x_i\in M_iX$, $y_i\in M_iY$ and $f_i:M_i{\restriction}L\xrightarrow{\cong}N{\restriction}L$ such that $M_i\models q[x_i,y_i]$, $f_{iY}y_i=y'$ and $\lambda_{M_1,N}[f_1]x_1=x'$, $\lambda_{M_2,N}[f_2]x_2=x''$; moreover, we may assume that M_1, M_2 are special. Since (M_1,x_1,y_1), (M_2,x_2,y_2) are both special models of the complete theory $q(\boldsymbol{x},\boldsymbol{y})$, they are isomorphic; say, by $h:(M_1,x_1,y_1)\xrightarrow{\cong}(M_2,x_2,y_2)$. Consider $g=f_2\circ(h|L):M_1{\restriction}L\xrightarrow{\cong}N{\restriction}L$; then the structures (M_2,N,f_2) and (M_1,N,g) of type L'' are isomorphic by the isomorphism $(h,1_N)$, hence

$$x''=(\lambda_{M_2,N}[f_2])x_2=(\lambda_{M_1,N}[g])h^{-1}x_2=(\lambda_{M_1,N}[g])x_1.$$

Also, $h^{-1}y_2=y_1$. What we have tells us that we may replace (M_2,y_2,x_2,f_2) by (M_1,y_1,x_1,g).

Now, also drop the subscript 1 since it is not necessary. We have the model M of $\mathcal{T}$, elements $x \in MX$ and $y \in MY$, and isomorphisms $M\restriction L \overset{f}{\underset{g}{\underset{\cong}{\rightrightarrows}}} N\restriction L$ such that $f_Y y = y'$, $\lambda[f]x = x'$ and $\lambda[g]x = x''$. By the support property (**), $x' = x''$ as required.

(21) implies the existence of a finite sub-conjunction $r(\mathbf{y}', \mathbf{x}')$ of $R(\mathbf{y}', \mathbf{x}')$ such that

$$\models \forall \mathbf{y}'\mathbf{x}'\mathbf{x}''((r(\mathbf{y}', \mathbf{x}') \wedge r(\mathbf{y}', \mathbf{x}'')) \longrightarrow \mathbf{x}' = \mathbf{x}'') .$$

r chosen in this way will obviously satisfy (*). [] 3

10. LIFTING ZAWADOWSKI'S ARGUMENT TO ULTRA*MORPHISMS

The category $\mathrm{Mod}(\mathcal{T}/L)$ "of $\mathcal{T}$-models with L-mappings" (see Section 3) carries an ultrastructure, consisting of ultraproducts and ultramorphisms; this structure is simply the restriction of the same structure on the category $\mathrm{Str}(L)$ of all L-structures along the full forgetful embedding of $\mathrm{Mod}(\mathcal{T}/L)$ into $\mathrm{Str}(L)$ (denoted F in Section 3). Similarly for $\mathrm{Mod}^*(\mathcal{T}/L)$. Now, each one of Zawadowski's descent data defines a functor $\mathrm{Mod}(\mathcal{T}/L) \longrightarrow \mathrm{Set}$; similarly, each IFS (see the previous section) defines one of the form $\mathrm{Mod}^*(\mathcal{T}/L) \longrightarrow \mathrm{Set}^*$. It is natural to ask whether such definable functors preserve the ultrastructure. One might first feel that this should be *obviously true*; after all, definable functors are "canonical enough", and canonical functors are usually expected to have good preservation properties. Zawadowski learned through hard work that this is not the case, and even that the preservation can be shown to hold only if the ultramorphisms are restricted to the *special* ones, the special ones that are constructed in [M1] (and made explicit in [M2]). This fact is, for me, the most surprising aspect of Zawadowski's work.

There is an intriguing self-referential element in Zawadowski's argument. The argument turns a failure of a definable functor preserving a special ultramorphism into a counter-example to the functoriality assumption on the definable functor, and to obtain the needed arrows between models, it uses the *same* construction as the one that went into the special ultramorphism itself, although on different parameters.

In this section, we go through the natural counterpart of Zawadowski's argument in the "Boolean" context.

In this section, we have the fixed theory $\mathcal{T} = (L', T)$ and the sublanguage $L \subset L'$ as in the previous section. We make the further mild assumption that L', L have only sorts and unary (sorted) operation symbols, no other operations, and no relations; this again will hold true for internal theories of pretoposes. In addition, we also fix $(X, \lambda[\mathbf{f}])$, a $\mathcal{T}/L$-IFS.

(i) We have the forgetful functor $F:\mathrm{Mod}^*(\mathcal{T}/L) \longrightarrow \mathrm{Str}^*(L)$ as in Sections 3 and 9. $\mathrm{Str}^*(L)$ is made into a Δ-u.gr. $\mathbf{Str}^*(L)$, with Δ the class of special u*.m.'s on Set, such that the family $\langle \mathrm{ev}_Y:\mathrm{Str}^*(L) \longrightarrow \mathrm{Set}^* \rangle_{Y\in \mathcal{S}_L}$ of evaluation functors ev_Y (for which $\mathrm{ev}_Y(M) = M(Y)$ and $\mathrm{ev}_Y(h:M \longrightarrow M') = h_Y:M(Y) \longrightarrow M(Y'))$ is a relation-conservative (see Section 4) family

$$\langle \mathrm{ev}_Y : \mathbf{Str}^*(L) \longrightarrow \mathbf{Set}^* \rangle_{Y \in \mathcal{S}_L}$$

of strict u*.f.'s (we drop reference to Δ); these specifications obviously define the ultragroupoid $\mathbf{Str}^*(L)$.

Further, we can make $\mathrm{Mod}^*(\mathcal{T}/L)$ uniquely into a u.gr. $\mathbf{Mod}^*(\mathcal{T}/L)$ so that the functor $F:\mathrm{Mod}^*(\mathcal{T}/L) \longrightarrow \mathrm{Str}^*(L)$ becomes a strict relation-conservative u*.f. $F:\mathbf{Mod}^*(\mathcal{T}/L) \longrightarrow \mathbf{Str}^*(L)$.

(ii) Let P be an arbitrary model of $\mathcal{T}$, and let $x \in PX$. By 9.3, with $p(\boldsymbol{x})$ the type of x in P, let $Y \in \mathcal{S}_L$, and $s(\boldsymbol{x}, \boldsymbol{y})$ be as there; let $y \in PY$ be such that $P \models s[x, y]$; let $q(\boldsymbol{x}, \boldsymbol{y})$ be the type of (x, y) in P; and let $r(\boldsymbol{y}, \boldsymbol{x})$ be an L'-formula for which (*) in 9.3 holds; we have the following:

(*) for any $M, N \models T$, $f:M\restriction L \xrightarrow{\cong} N\restriction L$, any $e:P \longrightarrow M$ in $\mathrm{Mod}(\mathcal{T})$ (elementary embedding!),

$$\sigma_{(M,N)}[f](ex) = x' \iff N \models r[f_Y ey, x'] .$$

(iii) Let $Z:\mathrm{Mod}^*(\mathcal{T}/L) \longrightarrow \mathrm{Set}^*$ be the functor defined by (X, σ); we have $ZM = MX$, and $Z((M, N, h:M|L \xrightarrow{\cong} N|L) = \sigma_{(M,N)}[h]$ $(M, N \models T)$. Z is a strict p.-u*.f. $Z:\mathbf{Mod}^*(\mathcal{T}/L) \longrightarrow \mathbf{Set}^*$. This is immediate, except for the preservation of the E-relations. We want to see the following: given

$$P \xrightarrow{\ e\ } M, \quad M\restriction L \underset{\cong\, g}{\overset{f\,\cong}{\rightrightarrows}} N\restriction L ,$$

(in fact, $P \xrightarrow{e} M$ of the form $P \xrightarrow{\ \delta\ } P^{\vec{U}, U}$, but we won't use this) with $f \circ (e\restriction L) = g \circ (e\restriction L)$, then $\sigma[f](e_X x) = \sigma[g](e_X x)$ for any $x \in PX$. Choose y for any given x as in (ii); the desired conclusion is clear from (ii)(*).

L', T, L, X, σ, Z, Y, s, r in (ii) and (iii) are all fixed for the rest of this section.

(iv) Let us have:

the rooted cell-system $G:\Lambda \longrightarrow \Gamma$;

$$\mathcal{M}:\Gamma \longrightarrow \mathbf{Mod}^*(\mathcal{T}/L)\ ;$$

$$P \underset{\mathrm{def}}{=} \mathcal{M}G\lambda_{\mathrm{in}},\ x\in PX,\ y\in PY;$$

with $\mathrm{ev}_Y:\mathbf{Mod}^*(\mathcal{T}/L)\longrightarrow \mathbf{Set}^*$ the evaluation functor at Y, a strict $\overset{*}{\mathrm{u}}$.f., a filling Φ of Λ along $\mathrm{ev}_Y\circ\mathcal{M}:\Gamma\longrightarrow\mathbf{Set}^*$ starting with y : $\Phi\lambda_{\mathrm{in}} = y$;

a filling Ξ of Λ along $Z\circ\mathcal{M}:\Gamma\longrightarrow\mathbf{Set}^*$ starting with x : $\Xi\lambda_{\mathrm{in}} = x$.

Consider the regular set $\Psi_0 = \mathrm{dom}(\Phi)\wedge\mathrm{dom}(\Xi)$. For any $\lambda\in\Psi_0$, we have $\Phi\lambda\in(\mathcal{M}G\lambda)Y$, $\Xi\lambda\in(\mathcal{M}G\lambda)X$. $\lambda\in\Lambda$ is a *point of matching* (relative to the system of data presupposed here) if $\lambda\in\Psi_0$ and $\mathcal{M}G\lambda\models r[\Phi\lambda,\Xi\lambda]$; the *matching set* is the set of all points of matching.

(v) Let $\vec{S} = \langle S_n\rangle_{n<\omega}$ be an isomorphism ultrascheme. Let m be a fixed natural number; we will apply the definition in (iv) for the cell-system involved in S_{2m}; therefore, we use the notation $S = S_{2m} = (G:\Lambda\longrightarrow\Gamma,\ \varphi,\ J,\ W,\ \pi)$. $\vec{S}$ defines the $\overset{*}{\mathrm{u}}$.m. $v = v^{\vec{S}}$ according to 7.(vi); we use the notation attached to $\vec{S}$ introduced in 7.(vi). With $\boldsymbol{K} \underset{\mathrm{def}}{=} \mathbf{Mod}^*(\mathcal{T}/L)$, we have the lifting $v_{\boldsymbol{K}}$, an $\overset{*}{\mathrm{u}}$.m. on $\boldsymbol{K}$, of $v = v_{\mathbf{Set}}$.

Let $\mathcal{M}:\Gamma_\omega\longrightarrow\mathbf{Set}^*$ be an $\overset{*}{\mathrm{u}}$.d. which is strict with respect to u^+, ℓ_ω, ℓ'_ω (see 7.(vi)). We use the same symbol $\mathcal{M}$ to denote the unique extension of $\mathcal{M}$ to Γ^ω strict w.r.to the additional specifications, as well as the various restrictions of this extensions to the Γ_n's.

Let us abbreviate $\mathcal{M}\gamma_{2n}$ as P_n ; then $P_n = P_0^{W_1,W_3,\ldots,W_{2n-1}}$ $(1\le n<\omega)$. Also, $P'_n = \mathcal{M}\gamma_{2n+1}$; then $P'_n = P_1^{W_2,W_4,\ldots,W_{2n}}$. We write P for P_m, and P' for P'_m ; then $P = \mathcal{M}G\lambda^S_{\mathrm{in}}$ in accordance with the notation in (iv).

When we evaluate $v_{\boldsymbol{K}}$ at $\mathcal{M}$, we obtain a function

$$v_{\boldsymbol{K},\mathcal{M}} : P_0^{\vec{W}} \longrightarrow P_0'^{\vec{W}'} .$$

Let us compare $Z(v_{\boldsymbol{K},\mathcal{M}})$ and $v_{\mathbf{Set},Z\circ\mathcal{M}} = v_{Z\circ\mathcal{M}}$. Both are functions of the form

$$(P_0X)^{\vec{W}} \rightrightarrows (P'_0X)^{\vec{W}'} .$$

Consider the diagonal map $\delta = \delta_{PX}^{\vec{W}, m} : PX \longrightarrow (P_0X)^{\vec{W}}$, and let $x \in PX$. We claim that

$$Z(\nu_{\mathbf{K}, \mathcal{M}})(\delta x) = \nu_{Z\circ\mathcal{M}}(\delta x) \tag{1}$$

provided that the set

$$Q \underset{\text{def}}{=} \{t \in J : \varphi\pi t \text{ is a point of matching}\} \in W.$$

To prove this, let us compute the right-hand-side first. With

$$\delta' = \delta_{P'}^{\vec{W}', m} : P' \longrightarrow P_0'^{\vec{W}'}$$

the diagonal map, we have, according to 7.(vi), that

$$\nu_{Z\circ\mathcal{M}}(\delta x) = \delta'(\mu_{Z\circ\mathcal{M}}^S(x));$$

and the definition of $\mu_{Z\circ\mathcal{M}} = \mu_{Z\circ\mathcal{M}}^S$ (see 7.(v)) gives

$$\mu_{Z\circ\mathcal{M}}(x) = \langle \Xi\varphi\pi t\rangle_{t\in Q}/W. \tag{2}$$

(Q being in W, we may use it as the range of the parameter t).

Turning to the left-side of (1), and with an eye on (ii)(*) in which we will consider $e = \delta$ and $f = \nu_{\mathbf{K},\mathcal{M}}$, we have

$$\nu_{\mathbf{K}, \mathcal{M}, Y}(\delta y) = \nu_{\text{ev}_Y\circ\mathcal{M}}(\delta y) = \delta'(\mu_{\text{ev}_Y\circ\mathcal{M}}(y))$$

and

$$\mu_{\text{ev}_Y\circ\mathcal{M}}(y) = \langle \Phi\varphi\pi t\rangle_{t\in Q}/W.$$

To show (1), by using (ii)(*) with x' the right-hand-side value in (2), it suffices to have

$$P'^{\vec{W}'} \models r[\delta'(\langle\Phi\varphi\pi t\rangle_{t\in Q}/W), \delta'(\langle\Xi\varphi\pi t\rangle_{t\in Q}/W),$$

which is equivalent to

$$P' \models r[\langle\Phi\varphi\pi t\rangle_{t\in Q}/W, \langle\Xi\varphi\pi t\rangle_{t\in Q}/W].$$

But by the definition of Q, this latter is clear (by Los's theorem).

(vi) Let us return to the situation of (iv), and let us make some obvious inferences about the matching set. Recall the definition of *closed* subset of Λ in 6.(ix). We claim that the matching set satisfies all clauses of the definition of "closed set" except possibly 6.(ix)(1).

6.(ix)(0) means that $P \models r[y, x]$. But this holds, by applying (ii)(*) for $M = N = P$, $e = 1_P$, $f = 1_{M\upharpoonright L}$, and $x' = x$; the reason is that the functor Z maps an identity into an identity.

The rest of the conditions follow directly from the definitions, in fact, independently of the fact how the fillings Φ, Ξ are defined. Indeed, let $u = (\lambda, I, U, Q, g) \in \mathcal{U}_2 \cup \mathcal{U}_3$. 6.(ix)(2) and 6.(ix)(3) together mean that $\mathcal{M}G\lambda \models r[\Phi\lambda, \Xi\lambda]$ iff there is $R \subset Q$, $R \in U$ such that $\mathcal{M}G\lambda i \models r[\Phi g(i), \Xi g(i)]$ for all $i \in R$. But this holds by Los's theorem, since $[\mathcal{M}, u] : \mathcal{M}G\lambda \xrightarrow{\cong} \prod_{i\in I} \mathcal{M}G\lambda i/U$, and in the filling Φ the elements $\Phi\lambda$ and the $\Phi g(i)$ are appropriately related as well as similarly for Ξ. The arguments for 6.(ix)(4) and 6.(ix)(5) are similar.

Note that for any fixed regular set Ψ, the set of points of matching in Ψ also satisfies the same conditions as in the claim.

(vii) Once again we return to the situation of (iv). We claim that the matching set contains a regular subset. Suppose the contrary; in other words, every regular set contains a point where matching fails. Let Ψ be an arbitrary regular set such that $\Psi \subset \Psi_0$. We claim that there is a *bad point* in Ψ, that is, a node $\lambda \in \Psi \cap \Lambda_1$ which is not a point of matching, but such that $\mu_{(1)\lambda}$ (necessarily in Ψ) is a point of matching. Indeed, if this were not the case, then, the set of points matching would satisfy condition (1) of the definition of "closed"; thus, by (vi), it would be a closed set, containing a regular set (see 6.(ix)), contrary to the assumption.

Let, for any $\Psi \in Q \underset{\text{def}}{=} \text{Reg}(\Lambda)$, λ_Ψ be a bad point in $\Psi \wedge \Psi_0$, and let φ be the function on Q for which $\varphi(\Psi) = \mu_{(1),\lambda_\Psi}$. Consider the ultrascheme $S = (\Lambda, \varphi, Q, W, 1_Q)$, with W any ultrafilter containing all sets of the form $[\Psi]$ (see 6.(xii)). Let $\vec{S}$ be any isomorphism extension of S; see 6.(xviii); we will use (v) with $m = 0$. Let v be the u.m. $v^{\vec{S}}$ defined by $\vec{S}$, and let $f : P^{\vec{W}} \xrightarrow{\cong} P'^{\vec{W}'}$ be the instance of $v_{\boldsymbol{K}}$ at $\mathcal{M}$; $f = v_{\boldsymbol{K},\mathcal{M}}$; here we used the notation introduced in (v) above.

On the other hand, we have $g_\Psi \underset{\text{def}}{=} \mathcal{M}(Ge_{\lambda_\Psi}) : \mathcal{M}G\varphi\Psi \xrightarrow{\cong} \mathcal{M}(G\lambda_\Psi)$, $g \underset{\text{def}}{=} \prod_{\Psi\in Q} g_\Psi/W : P' \longrightarrow P''$, with $P'' \underset{\text{def}}{=} \prod_{\Psi\in Q} \mathcal{M}(G\lambda_\Psi)/W$, and $h \underset{\text{def}}{=} g^{\vec{W}'} : P'^{\vec{W}'} \xrightarrow{\cong} P''^{\vec{W}'}$. Let us apply the functor Z to the composite $h \circ f : P^{\vec{W}} \xrightarrow{\cong} P''^{\vec{W}'}$, and let us calculate the effect of the function

$$Z(h\circ f) : (PX)^{\vec{W}} \longrightarrow (P''X)^{\vec{W}'}$$

on the element $\delta(x) \in (PX)^{\vec{W}}$. For one thing, $Z(h \circ f) = (Zh) \circ (Zf)$. For another, $Zh = (\prod_{\Psi \in Q} (Zg_\Psi)/W)^{\vec{W}}$; this follows directly from the definition of Z as a functor specified by (X, σ), essentially because the ultraproduct $\prod_{i \in I} (M_i, N_i, h_i : M_i \restriction L \xrightarrow{\cong} N_i \restriction L)/U$ is the same as $(\prod_{i \in I} M_i/U, \prod_{i \in I} N_i/U, \prod_{i \in I} h_i/U)$, and similarly for limit ultrapowers.

By (v), since $\varphi\Psi$ is a point of matching for every $\Psi \in Q$, we have

$$(Zf)(\delta x) = \delta'(\langle \Xi\varphi\Psi \rangle_{\Psi \in Q}/W),$$

and thus

$$Z(h \circ f)(\delta x) = (Zh)(Zf)(\delta x) = \delta'(\langle (Zg_\Psi)(\Xi\varphi\Psi) \rangle_{\Psi \in Q}/W).$$

Now, let us apply (ii)(*) with $h \circ f$ for f, δ for e, and the right-hand-side value of the last formula for x'. We get

$$P''^{\vec{W}'} \models r[h_Y f_Y \delta y, x'].$$

Here, by (v), $f_Y \delta y = \delta'(\langle \Phi\varphi\Psi \rangle_{\Psi \in Q}/W)$, and thus

$$h_Y f_Y \delta y = \delta'(\langle g_\Psi(\Phi\varphi\Psi) \rangle_{\Psi \in Q}/W.$$

Therefore, we obtain

$$P'' \models r[\langle g_\Psi(\Phi\varphi\Psi) \rangle_{\Psi \in Q}/W, \langle (Zg_\Psi)(\Xi\varphi\Psi) \rangle_{\Psi \in Q}/W],$$

and

$$\{\Psi \in Q : \mathcal{M}(G\lambda_\Psi) \models r[g_\Psi(\Phi\varphi\Psi), (Zg_\Psi)(\Xi\varphi\Psi)]\} \in W.$$

But by the definition of the fillings Φ, Ξ, and by the meaning of g_Ψ, we have that $\Phi\lambda_\Psi = g_\Psi(\Phi\varphi\Psi)$, $\Xi\lambda_\Psi = (Zg_\Psi)(\Xi\varphi\Psi)$. Thus,

$$\{\Psi \in Q : \mathcal{M}(G\lambda_\Psi) \models r[\Phi\lambda_\Psi, \Xi\lambda_\Psi]\} \in W.$$

This means that λ_Ψ is a point of matching for a set of Ψ's in W; but this contradicts the choice of λ_Ψ as never being a point of matching for any Ψ !

(viii)(conclusion) Any definable functor $Z:\mathbf{Mod}^*(\mathcal{T}/L) \longrightarrow \mathbf{Set}^*$ preserves the special ultramorphisms.

Indeed, the claim of (v) (see (1)) says that the definable functor Z carries the u.*m. ν_K

on $\boldsymbol{K} = \mathbf{Mod}^*(\mathcal{T}/L)$ to the original ν on $\mathtt{Set}$, at least as far sufficiently strict $\overset{*}{\mathrm{u.}}$d.'s $\mathcal{M}$ are concerned, *provided* that there are enough points of matching in the relevant situation. But (vii) says that the last condition will always be fulfilled. Thus, the preservation of ν is assured as far as sufficiently strict $\overset{*}{\mathrm{u.}}$d.'s are concerned. In fact, the general case of an arbitrary $\mathcal{M}$ follows directly from the special case, via the reduction already used at the end of 7.(vi).

11. THE OPERATIONS IN $\mathcal{BP}^*$ AND UG

$\mathcal{CAT}$, the 2-category of categories, functors and natural transformations, is bicomplete. Moreover, it has a stronger property; for instance, in the definition of "comma square", the functor that is required to be an equivalence can be made, by the natural choice of the limit as indicated in Section 2, to be an isomorphism of categories. This has the consequence that all the isomorphisms ι_0, ι_1, ρ_0, ρ_1, ρ_2, become identities. We talk about the *strong* bipullback, or more generally, *strong* bilimits when the stronger definition is intended.

$\mathcal{BP}^*$ has an obvious forgetful functor to $\mathcal{CAT}$, and this creates all strong bilimits in $\mathcal{BP}^*$; in other words, all strong bilimits are calculated in $\mathcal{BP}^*$ as in $\mathcal{CAT}$. Let us note that *ordinary* limits such as pullbacks do not even exist in $\mathcal{BP}^*$ (and in related 2-categories of structured categories), let alone having desired meanings. We refer to [B/K/P] for these matters.

In $\mathcal{BP}^*$, we are interested in the bilimit which is the co-quotient of a co-c-complex; in fact, we will need it when the co-c-complex is the cokernel complex of an arrow I, in which case we will call the co-quotient the *descent object* of I, and denote it by $\mathrm{Des}(I)$. Let the diagram

$$T \underset{P_1}{\overset{P_0}{\underset{\longrightarrow}{\overset{\longrightarrow}{\xleftarrow{\;\Delta\;}}}}} T_1 \underset{P_{12}}{\overset{P_{01}}{\underset{\longrightarrow}{\overset{\longrightarrow}{\xrightarrow{\;P_{02}\;}}}}} T_2 \,, \tag{1}$$

denoted by $\mathcal{D}$, be a *strict* co-c-complex, that is, we have the duals of the five equalities in the definition of c-complex, with the isomorphisms all made equal to identities. The co-quotient of $\mathcal{D}$, that is, the universal pair $(H:D\to T,\ \delta:P_0H\xrightarrow{\cong}P_1H)$ with $\Delta\delta = 1_H$, $P_{02}\delta = P_{12}\delta\circ P_{01}\delta$, is given as follows. The category D has

objects the pairs $(X\in T,\ \xi:P_0X\xrightarrow{\cong}P_1X)$ for which $\Delta\xi = 1_X$ and $(P_{12}\xi)\circ(P_{01}\xi) = P_{02}\xi$, and

morphisms $(X,\xi)\to(X',\xi')$ the morphisms $x:X\to X'$ in T such that

$$\begin{array}{ccc} P_0X & \xrightarrow{\xi} & P_1X \\ {\scriptstyle P_0x}\downarrow & & \downarrow{\scriptstyle P_1x} \\ P_0X' & \xrightarrow{\xi'} & P_1X' \end{array}$$ commutes; composition is the obvious one.

The functor H takes (X, ξ) to X, and $x:(X, \xi)\to(X', \xi')$ to $x:X\to X'$. Finally, $\delta_{(X, \xi)} = \xi$.

It is easy to verify the fact that the described items give the co-quotient in $\mathcal{CAT}$; the same follows for $\mathcal{BP}^*$ from what we said above.

We make some general remarks concerning bilimits, in particular, the co-quotient construction, in $\mathcal{CAT}$. The first is the fact of invariance under equivalences. Suppose $\mathcal{D}$ and $\mathcal{D}'$ are both strict co-c-complexes, $\mathcal{D}$ in the notation in (1), $\mathcal{D}'$ similarly with all items primed. Let $\vec{J}:\mathcal{D}\longrightarrow\mathcal{D}'$ be an arrow (natural transformation), that is $\vec{J} = (J_0, J_1, J_2)$, $J_i:T_i\to T'_i$ ($T_0 = T$, ...), with six commutative squares, corresponding to the six arrows in each complex. With the notation for the co-quotient of $\mathcal{D}$ given above, and with the primed version used for $\mathcal{D}'$, we have, as a consequence of the universal property of (H', δ'), a (unique) $J:D\to D'$ such that $J_0H = H'J$, $J_1\delta = \delta' J$. We claim that if each J_i $(i=0, 1, 2)$ is an equivalence of categories, so is J. This is a case of a very general fact about bilimits; it is easy to see it directly as follows.

Let's show that J is essentially surjective. Let $(X', \xi')\in D'$. Then, since J_0 is an equivalence, there is $X\in T$ with an isomorphism $x':X'\xrightarrow{\cong}J_0X$. Define $\xi'':P'_0J_0X\xrightarrow{\cong}P'_1J_0X$ as $\xi'' = (P'_1x')\circ\xi'\circ(P'_0x')$. $(J_0X, \xi'')\in D'$ as a consequence of $(X', \xi')\in D'$. Since J_1 is full and faithful, there is $\xi:P_0X\xrightarrow{\cong}P_1X$ such that $J_1\xi = \xi''$ (note $P'_0J_0 = J_1P_0$, ...). By the faithfulness of $\vec{J}$, and since the $\vec{J}$-image of (X, ξ) is in D', we have $(X, \xi)\in D$. Also, $J((X, \xi))\cong(X', \xi')$ via x'.

Another point that will be of use for us concerns the very definition of "quotient". Suppose $\mathcal{C}$ is c-complex, strict for simplicity, in any 2-category; let us assume the notation 2.(1) for $\mathcal{C}$; let B be any object. The category $\mathrm{Equ}_{\mathcal{C}}(B)$ may be regarded as a limit, in fact, co-quotient, in $\mathcal{CAT}$, in a natural way as follows. One applies the contravariant 2-functor $\mathrm{Hom}(-, B)$ to $\mathcal{CAT}$ to obtain the co-c-complex

$$\mathrm{Hom}(E, B) \overleftarrow{\overleftarrow{\overleftarrow{\quad}}} \mathrm{Hom}(K, B) \overleftarrow{\overrightarrow{\quad}} \mathrm{Hom}(A, B) \quad . \tag{2}$$

The co-quotient of this co-complex, calculated in $\mathcal{CAT}$ in the standard way explained above, is exactly $\mathrm{Equ}_{\mathcal{C}}(B)$.

Let us turn to bicolimits in $\mathcal{BP}^*$; these are more complicated.

Let $I: S \longrightarrow T$ be an arrow in $\mathcal{BP}^*$. Let us make the very mild restriction on I that it be injective on objects: $I(A) = I(B)$ cannot happen for distinct objects A and B of S. It is easy to see that any $I: S \longrightarrow T$ is *equivalent* to one with the mentioned property; since our final conclusions are invariant under equivalence, we have not lost generality. (Of course, the mentioned restriction is in no way essential, not even in a technical sense. Without it, in Sections 3, 9 and 10, we should consider a not necessarily injective map $L \longrightarrow L'$ between languages, instead of a sublanguage L of L'. In definability theory, it is customary to talk about a situation involving a language and a sublanguage of it, and it seemed a good idea to follow that practice here.)

We let $\mathcal{T} = (L', T)$ be the internal theory of T, and $L \subset L'$ the image of I. In Section 9, we defined the theory $\mathcal{T}+_L\mathcal{T}$. We have the *pretopos completion* $\mathcal{P}r(\mathcal{T}+_L\mathcal{T})$ of $\mathcal{T}+_L\mathcal{T}$ described in detail in [M/R], Section 8.4. In the cocomma square

$$\begin{array}{ccccc} & & T & & \\ & \overset{I}{\nearrow} & \mu\Big\downarrow\cong & \overset{P_0}{\searrow} & \\ S & & & & T+_S T\ , \\ & \underset{I}{\searrow} & & \underset{P_1}{\nearrow} & \\ & & T & & \end{array} \qquad (3)$$

$T+_S T$ is (can be taken to be) $\mathcal{P}r(\mathcal{T}+_L\mathcal{T})$.

The construction of the pretopos completion $\mathcal{P}r(\mathcal{S})$ of any theory $\mathcal{S}$ proceeds in two steps. In the first, one constructs the *coherent completion*, or Lindenbaum-Tarski category $\mathcal{C}oh(\mathcal{S})$ of $\mathcal{S}$; both the objects and the morphisms of $\mathcal{C}oh(\mathcal{S})$ are given by formulas of the theory $\mathcal{S}$; see [M/R], 8.2, or [M3], 2.5. In the second step, one takes the *(free) pretopos completion* $\mathcal{P}r(\mathcal{C}oh(\mathcal{S}))$ of $\mathcal{C}oh(\mathcal{S})$; a purely categorical description of $\mathcal{P}r(\mathbf{C})$, with $\mathbf{C}$ a coherent category, is described in section 2.4 of [M3]. $\mathbf{C}$ is a full and faithful subcategory of $\mathcal{P}r(\mathbf{C})$; as we will see, this fact makes it possible to ignore the particulars of the description of $\mathcal{P}r(\mathbf{C})$.

For every formula $\varphi(\vec{x})$ of $\mathcal{S}$, there is an object $[\vec{x}:\varphi]$ of $\mathcal{C}oh(\mathcal{S})$; the functor $P_0: T \to T+_S T$ maps, by definition, the object X of T to the object $[x: X_{(0)}(x)]$ of $T+_S T$; similarly for P_1; there is a similar straightforward description of the effect of P_0, P_1 on arrows. The diagram (3) so described is indeed a cocomma square, in fact, in such a way that the equivalence functor figuring in the description of the universal property is a *surjective equivalence*, that is, it is strictly surjective on objects. The proof is easy on the basis of the universal properties of the coherent and pretopos completions given in *loc.cit.*

Similarly, the final object $T+_S T+_S T$ in the cokernel complex

$$T \xleftarrow{\Delta} T+_S T \quad (P_0, P_1 \colon T+_S T \to T) \qquad T+_S T \xrightarrow{P_{01},\ P_{02},\ P_{12}} T+_S T+_S T \tag{4}$$

of I is the pretopos completion of the theory $\mathcal{T}+_L\mathcal{T}+_L\mathcal{T}$ described in Section 9., with P_{01} and P_{12} given in the straightforward manner; in fact, the isomorphism $\rho_1 : P_1P_{01} \xrightarrow{\cong} P_0P_{12}$ can also be chosen to be the identity. Because of the surjective quality of the universality-equivalences, in the construction of Δ and P the additional isomorphisms ι_0, ι_1, ρ_0, ρ_2 turn out to be identities as well. Thus, in (4) we have a strict co-c-complex.

Looking at the construction of the co-quotient given above, we see that it uses only the coherent completions $Coh(\mathcal{T}+_L\mathcal{T})$, $Coh(\mathcal{T}+_L\mathcal{T}+_L\mathcal{T})$, which are full subcategories of the corresponding pretoposes. An arrow in $Coh(\mathcal{T}+_L\mathcal{T})$ is given by a formula that is provably functional with the appropriate domain and codomain, with provably equivalent formulas giving the same arrows. The notation is $[\vec{x}\longmapsto\vec{y}:\mu]$ for an arrow $[\vec{x}:\varphi]\longrightarrow[\vec{y}:\psi]$, with the formula having the free variables $\vec{x}$, $\vec{y}$ at most, and satisfying the functionality condition. Therefore, the arrow $\xi : P_0X \longrightarrow P_1X$ in an object $(X, \xi) \in \mathrm{Des}(I)$, that is, a set of "descent data", is given by a formula $\lambda[\boldsymbol{f}](\boldsymbol{x},\boldsymbol{x}')$ with $(X,\lambda[\boldsymbol{f}])$ an IFS as defined in Section 9; ξ is $[\boldsymbol{x}\longmapsto\boldsymbol{x}':\lambda[\boldsymbol{f}]] : [\boldsymbol{x}:X_{(0)}\boldsymbol{x}] \longrightarrow [\boldsymbol{x}':X_{(1)}\boldsymbol{x}']$; conditions (9), (10) in 9. correspond to the two equalities required above for an object of the co-quotient.

Let us turn to the exactness properties of **UG** . In this 2-category, we are interested in the kernel complex of an arrow, and the quotient of the latter. We try to relate as much as possible of these structures of **UG** to $\mathcal{GPD}$ which is the full sub-2-category of $\mathcal{CAT}$ on the groupoids as objects. Note that all 2-arrows in $\mathcal{GPD}$ are isomorphisms.

The computation of a kernel complex in $\mathcal{GPD}$ is as in $\mathcal{CAT}$; the construction was given (in pieces) in Section 2. Note that now the comma-square becomes the same as the bipullback. We noted that in $\mathcal{CAT}$ the quotient of the kernel complex of an arrow gives the e.s./f.f. factorization, and may be arranged so that the universality equivalence is in fact an isomorphism of categories. The computation of the quotient in $\mathcal{GPD}$ is as in $\mathcal{CAT}$.

A *pre*$^-$*-ultragroupoid* is a "pre-ultragroupoid minus the E-relations"; it is a category with ultraproduct and ultralimit functors. The notion of *pre*$^-$*-ultra***functor* is clear; the notion of ultra*transformation need not be changed; we now have the 2-category $\mathbf{UG}^-$ of these things. The point is that everything transfers to $\mathbf{UG}^-$ from $\mathcal{GPD}$. First of all we have the forgetful functor $|-|:\mathbf{UG}^- \longrightarrow \mathcal{GPD}$, and $\mathbf{UG}^-$ has all strong bilimits preserved by $|-|$. Furthermore, the (2-)regular factorization in $\mathbf{UG}^-$ is just like in $\mathcal{GPD}$, and it agrees with the e.s./f.f. factorization. In more detail, if $X:K\longrightarrow L$ is an arrow in $\mathbf{UG}^-$, assumed a strict p$^-$-u*.f. for simplicity, then taking Q to be the category whose objects are those of K, and whose arrows $M\dot{\longrightarrow}N$ are the arrows $XM\longrightarrow XM$ in L, we can endow Q with a pre$^-$-ultrastructure in a unique way so that the canonical functors $K\to Q\to L$ both become strict p$^-$-u*.f.'s. Moreover, the just described (canonical choice for the) e.s./f.f. factorization answers the universal property, with a universality *isomorphism* of categories, of the regular factorization. Thus, the forgetful functor preserves "precisely" the regular and the e.s./f.f. factorizations.

Although these facts are, of course, fundamental, and their verification throws light on the "algebra" of factorizations in 2-categories, they are entirely formal and straightforward; we will not take up space with the details.

Turning to the "full" 2-category $\mathbf{UG}$, recall that we want to talk about *inclusions* in $\mathbf{UG}$. By definition, an u*.f. is an *inclusion* if its functor part is full and faithful, and it is relation conservative. (It is possible to define "inclusion" by an extremal property if we also bring in the forgetful $\mathbf{UG}\longrightarrow\mathbf{UG}^-$). Now, what is clear is that any arrow in $\mathbf{UG}$ can be factored into an arrow whose underlying functor is e.s., and one which is an inclusion; in fact, the factorization described above for $\mathbf{UG}^-$ gives the desired result; we only have to take the E-relations and the ultramorphisms on Q as given by L, in the obvious sense. What is not clear is that in this way we get the regular factorization; in fact, we are not claiming that we always do.

When we apply the factorization described in the last paragraph to $I^* = \mathbf{Mod}^*(I):\mathbf{Mod}^*(T)\longrightarrow\mathbf{Mod}^*(S)$, we get

$$\begin{array}{ccc} \mathbf{Mod}^*(T) & \xrightarrow{\qquad\qquad I^*\qquad\qquad} & \mathbf{Mod}^*(S)\ , \\ & {\scriptstyle R}\searrow \quad \mathbf{Mod}^*(T/S) \quad \nearrow{\scriptstyle F} & \end{array}$$

with $\mathbf{Mod}^*(T/S)$ the same as $\mathbf{Mod}^*(\mathcal{T}/L)$ at the beginning of Section 10. R is the functor that takes M to M, and $h:M\xrightarrow{\cong}N$ to $h\upharpoonright L$; F is defined in 10.(i).

An important connection between $\mathbf{UG}$ and $\mathbf{UG}^-$ is that if $K^-\overset{X}{\underset{Y}{\rightrightarrows}}K'^-$ are two p$^-$-u*.f.'s (arrows in $\mathbf{UG}^-$), with K^- and K'^- being the underlying p$^-$-u.gr.'s of the u.gr.'s K and K', respectively, moreover X is in fact an u*.f. (an arrow of $\mathbf{UG}$), and $Y\cong X$ in $\mathbf{UG}^-$, then Y is also an u*.f. In brief, being an u*.f. is invariant under (ultra)isomorphisms. The proof of this is easy; it uses the invariance property of the E-relations built into the concept of u.gr.

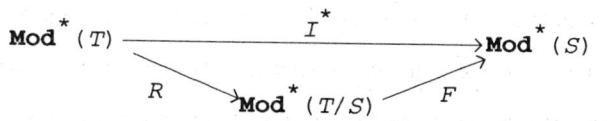

12. CONCLUSION

Let $I:S\to T\in \mathcal{BP}$, and consider the image of the co-complex 11.(4) under the left adjoint $F=(\ )^*:\mathbf{UG}\to\mathcal{BP}^{*\mathrm{op}}$ of the Stone-adjunction of Section 5, the complex $\mathcal{C}$:

$$(T+_S T+_S T)^* \underset{P_{12}^*}{\overset{P_{01}^*,\ P_{02}^*}{\longrightarrow}} (T+_S T)^* \underset{P_1^*}{\overset{P_0^*}{\underset{\Delta^*}{\rightleftarrows}}} T^* \ . \tag{1}$$

Since F is a left adjoint, and thus takes a colimit in $\mathbf{UG}$ into a limit in $\mathcal{BP}^*$, the last diagram is the kernel complex of I^*. The $\overset{*}{\mathrm{u}}$.f. R introduced in the last diagram of Section 11 comes equipped with the $\overset{*}{\mathrm{u}}$.t. $\rho:RP_0^*\xrightarrow{\cong}RP_1^*$ for which $\rho_{(M,\,N,\,h:M|L\xrightarrow{\cong}N|L)}=h$. We claim that

(i) *the pair* $q \underset{\mathrm{d\bar{e}f}}{=} (R,\rho)$ *satisfies the strong universal property for being the quotient of the c-complex* (1) *when tested with the specific object* $\mathbf{S}\underset{\mathrm{d\bar{e}f}}{=}\mathbf{Set}^*$ *in* $\mathbf{UG}$;

that is, $q_{\mathbf{S}}^*:\mathrm{Hom}(\boldsymbol{K},\mathbf{S})\to\mathrm{Equ}_{\mathcal{C}}^{(\mathbf{UG})}(\mathbf{S})$ (see Section 2) is an isomorphism of categories; $\boldsymbol{K}=\mathbf{Mod}^*(T/S)$. To see this, first note that the arrows and 2-arrows in $\mathbf{UG}$ are the same as those in $\mathbf{UG}^-$ except for the fact that the domains and codomains are restricted, and the arrows are required to satisfy an additional condition (preserving special $\overset{*}{\mathrm{u}}$.m.'s). In other words, $\mathrm{Hom}_{\mathbf{UG}}(\boldsymbol{K},\mathbf{S})$ and $\mathrm{Equ}_{\mathcal{C}}^{(\mathbf{UG})}(\mathbf{S})$ are full subcategories of $\mathrm{Hom}_{\mathbf{UG}^-}(\boldsymbol{K},\mathbf{S})$ and $\mathrm{Equ}_{\mathcal{C}}^{(\mathbf{UG}^-)}(\mathbf{S})$, respectively, and $q_{\mathbf{S}}^{*(\mathbf{UG}^-)}:\mathrm{Hom}_{\mathbf{UG}^-}(\boldsymbol{K},\mathbf{S})\xrightarrow{\cong}\mathrm{Equ}_{\mathcal{C}}^{(\mathbf{UG}^-)}(\mathbf{S})$ extends its own version for $\mathbf{UG}$.

It follows that it suffices to show that

(ii) *for any* $(Y,\zeta)\in\mathrm{Equ}_{\mathcal{C}}^{(\mathbf{UG})}(\mathbf{S})$, $(q_{\mathbf{S}}^{*(\mathbf{UG}^-)})^{-1}(Y,\zeta)$ *belongs to* $\mathrm{Hom}_{\mathbf{UG}}(\boldsymbol{K},\mathbf{S})$, *i.e. it is an ultra*functor.*

We recall that $\mathrm{Equ}_{\mathcal{C}}^{(\mathbf{UG})}(\mathbf{S})$ is the co-quotient of the co-c-complex in 11.(2), obtained from $\mathcal{C}$ by applying the functor $\mathrm{Hom}(-,\mathbf{S}):\mathbf{UG}^{\mathrm{op}}\longrightarrow\mathcal{CAT}$. But the system $\vec{\varepsilon}=(\varepsilon_T,\varepsilon_{T_1},\varepsilon_{T_2})$ is an arrow (natural transformation) from the co-c-complex 11.(4) to 11.(2); here $T_1=T+_ST$, $T_2=T+_ST+_ST$, and, e.g., ε_T is the canonical functor

$$\varepsilon_T : T\longrightarrow\mathrm{Hom}(T^*,\mathbf{S})$$

(now understood to be a mere functor, rather than a pretopos functor), which is shown in Theorem 8.1 to be an equivalence of categories. The naturality of $\vec{\varepsilon}$ is a consequence of the naturality of $\varepsilon:1_{\mathcal{BP}^{*\mathrm{op}}}\longrightarrow FG$. Thus, using "invariance under equivalence" from the last section, with $D=\mathrm{Des}(I)$ we have an induced map $\varepsilon:D\longrightarrow\mathrm{Equ}_{\mathcal{C}}^{(\mathbf{UG})}(\mathbf{S})$, and we have that it is an equivalence of categories.

The components of $\vec{\varepsilon}$ are evaluations, and consequently, ε is an "evaluation" too; inspection shows that for an object $(X,\xi)\in D$ given by the $\mathcal{T}/L$-IFS $(X,\lambda[\mathbf{f}])$, $\varepsilon(X,\lambda)$ is $(Y,\zeta)\in\mathrm{Equ}_{\mathcal{C}}^{(\mathbf{UG})}(\mathbf{S})$ where $Y=\varepsilon_T(X)\in\mathrm{Hom}(T^*,\mathbf{S})$, in particular

$$Y(M)\qquad = M(X)\qquad (M\in\mathbf{Mod}^*(T));$$

and

$$\zeta_{(M,N,f:M|L\xrightarrow{\cong}N|L)} = \lambda_{(M,N)}[f]\quad (M,N\in\mathbf{Mod}^*(T))$$

where we used the notation in Section 9. It follows (see Section 2 on how $(q_K^*)^{-1}$ acts) that $(q_{\mathbf{S}}^{*(\mathbf{UG}^-)})^{-1}(\varepsilon(X,\xi))$ is the same as the functor

$$[X,\lambda[\mathbf{f}]] : \mathbf{Mod}^*(\mathcal{T}/L)\longrightarrow\mathbf{Set}^*$$

specified by $(X,\lambda[\mathbf{f}])$ in the sense of Section 9. By 10.(viii), this is an u*.f. We have obtained that (ii) is true for (Y,ζ) in the image of ε. But ε is an equivalence, hence, essentially surjective on objects. Since being an u*.f. is invariant under isomorphisms (see the end of the last section), (ii) follows, whence (i) does too.

An u.gr. $\mathbf{K}$ is called *totally separated* (t.s.) if there is a family $\mathcal{X}=\langle X_i\rangle_{i\in I}$ of u*.f.'s $X_i:\mathbf{K}\longrightarrow\mathbf{Set}^*$ that is relation-conservative and 2-faithful (the latter meaning that the X_i induce a jointly one-to-one mapping on 2-arrows between two fixed arrows). Note that if $\mathbf{K}$ is t.s. witnessed by $\langle X_i\rangle_{i\in I}$ and $X:\mathbf{L}\longrightarrow\mathbf{K}$ is a pre$^-$-ultra*functor such that each $X_i\circ X$ is an ultra*functor, then X is an ultrafunctor itself. We claim that

(iii) (i) *holds with* $\mathbf{S}$ *replaced with any t.s.* $\boldsymbol{K}$.

As before, it suffices to have (ii) with $\boldsymbol{K}$ for $\mathbf{S}$. For $(Y, \zeta) \in \mathrm{Equ}_{\mathcal{C}}^{(\mathbf{UG})}(\boldsymbol{K})$, we have $(X_i Y, X_i \zeta) \in \mathrm{Equ}_{\mathcal{C}}^{(\mathbf{UG})}(\mathbf{S})$; hence $Z_i \underset{\mathrm{def}}{=} (q_{\mathbf{S}}^{*(\mathbf{UG}^-)})^{-1}(X_i Y, X_i \zeta)$ is an ultrafunctor for all i. Clearly, $Z_i = X_i \circ ((q_{\boldsymbol{K}}^{*(\mathbf{UG}^-)})^{-1}(Y, \zeta)$, and the assertion follows.

Let $\mathbf{UG}_{\mathrm{t.s.}}$ be the full sub-2-category of $\mathbf{UG}$ on the objects the totally separated ultragroupoids. $\mathbf{UG}_{\mathrm{t.s.}}$ is closed under bilimits in $\mathbf{UG}$ as easily seen. For any $T \in \mathcal{BP}^*$, $\mathbf{Mod}^*(T)$ is t.s., as witnessed by the evaluation functors (strict u.*f.'s) ev_A $(A \in T)$ (see 5.(i)*(b)). Thus, the Stone adjunction persists with $\mathbf{UG}_{\mathrm{t.s.}}$ in place of $\mathbf{UG}$. Also, $\mathbf{Mod}^*(T/S) = \mathbf{Mod}^*(\mathcal{T}/L)$ is in $\mathbf{UG}_{\mathrm{t.s.}}$ as shown by $F:\mathbf{Mod}^*(\mathcal{T}/L) \to \mathbf{Mod}^*(S)$ above, and by $\mathbf{Mod}^*(S) \in \mathbf{UG}_{\mathrm{t.s.}}$.

Theorem 2.4 is now proved; the Zawadowski setup for $\mathcal{BP}$ is

$$\mathcal{BP}^{*\mathrm{op}} \underset{\mathbf{Mod}^*(-)}{\overset{\mathcal{H}om(-,\mathbf{Set})}{\rightleftarrows}} \mathbf{UG}_{\mathrm{t.s.}};$$

condition Z1 is Gödel/Deligne completeness, Z2 is 8.2, and Z3 is (iii).

A theorem of Haim Gaifman's (see [M3]; "conceptual completeness") amounts to the following:

for a pretopos morphism $I:S \longrightarrow T$ *between small Boolean pretoposes* S *and* T, $I^*:\mathrm{Mod}^* T \longrightarrow \mathrm{Mod}^* S$ *is full and faithful if and only if* I *is a quotient morphism.*

Note that the "if" part is obvious. The theorem was stated (and proved) in [M3] as 3.1.4 for countable theories; in an addendum at the end of [M3], it was mentioned that one can infer the general case from the countable case.

Assume I^* is full and faithful. I^* is, trivially, a strict ultra*functor $\mathbf{Mod}^* T \longrightarrow \mathbf{Mod}^* S$. If it were relation-conservative, then the "only if" part of Gaifman's theorem would follow from 8.2 (and 8.1), with S for T, $\mathrm{Mod}^* T$ for $\boldsymbol{K}$, and I^* for Z; now, $Z^{\#}: S \longrightarrow \mathcal{H}om(\mathbf{Mod}^* T, \mathbf{Set})$ is $\varepsilon_T \circ I$. We do have that I^* is relation-conservative, under the assumption that it is full and faithful, as we proceed to show.

Consider the language $\hat{L}'$ of structures $(M, N, \langle h_X \rangle_{X \in T})$, with M and N structures for the underlying language of the internal theory $\mathcal{T}$ of T, and h_X an operation symbol

$h_X : X_{(0)} \longrightarrow X_{(1)}$. Let $\hat{L}$ be the sub-language of $\hat{L}'$ retaining only those h_X for which X is in the image of I; we rename h_{IY} as f_Y. Let the theory $\hat{T}$ over $\hat{L}'$ say that M, $N \models \mathcal{T}$, and that $\langle h_X \rangle_{X \in T}$ is an isomorphism $M \xrightarrow{\cong} N$. Note that the faithfulness of I^* says that for any $\hat{L}$-structure $P = (M, N, \langle f_Y \rangle_{Y \in S})$, there is at most one $\hat{L}'$-expansion of P which is a model of $\hat{T}$. By the Beth definability theorem, there are formulas $\lambda_X[\mathbf{f}](\mathbf{x}, \mathbf{x}')$ with $\mathbf{x} : X_{(0)}$, $\mathbf{x}' : X_{(1)}$ such that for any P as above, the unique $(\hat{L}', \hat{T})$-expansion $(M, N, \langle h_X \rangle_{X \in T})$, if exists, is given by $h_X = (\lambda_X)_{(M, N)}[f]$. The fullness of I^* says that for any M, $N \models \mathcal{T}$ and isomorphism $f = \langle f_Y \rangle_{Y \in S} : M \restriction L \xrightarrow{\cong} N \restriction L$, the $(\hat{L}', \hat{T})$-expansion $(M, N, \langle h_X \rangle_{X \in T})$ does exist. It follows that $(X, \lambda_X[\mathbf{f}])$ is a $\mathcal{T}/L$-isomorphism functor specification for all $X \in T$. By 10.(iii), the functor $Z_X : \mathrm{Mod}^*(\mathcal{T}/L) \longrightarrow \mathrm{Set}^*$ defined by $(X, \lambda_X[\mathbf{f}])$ preserves the E-relations. Clearly, for any $h : M \xrightarrow{\cong} N$, $h_X = Z_X(I^* h)$. Apply the last to both h and k in

$$M \xrightarrow{\delta} M^{\vec{U}} \overset{h}{\underset{k}{\rightrightarrows}} N$$

and

$$M|L \xrightarrow{\delta} M|L^{\vec{U}} \overset{I^* h}{\underset{I^* k}{\rightrightarrows}} N|L \, .$$

If $(I^* h) \circ \delta = (I^* k) \circ \delta$, then $Z_X((I^* h)) \circ \delta_X = Z_X((I^* k)) \circ \delta_X$, i.e. $h_X \circ \delta_X = k_X \circ \delta_X$ for all $X \in T$; that is, $h \circ \delta = k \circ \delta$. This shows that I^* is relation-conservative, and completes our proof of Gaifman's theorem .

Finally, let us indicate a "reduced" proof of 2.2, or, what is the same, 3.1. This proof takes place in the context of a pretopos morphism $I : S \to T$, and the induced functors

$$S \xrightarrow{J} \mathrm{Des}(I) \xrightarrow{K} \mathcal{H}om(\mathbf{Mod}^*(\mathcal{T}/L), \mathbf{Set}) \underset{\mathrm{def}}{=} R \, , \qquad (2)$$

and it aims at proving that

J is a quotient morphism. (3)

For the purposes of the present approach, it is best to *define* $\mathrm{Des}(I)$ as the category whose objects are the $\mathcal{T}/L$-IFS's , and whose arrows $(X, \lambda) \longrightarrow (X', \lambda')$ are arrows $x : X \longrightarrow X'$ in T such that for all $(M, N, h) \models \mathcal{T} +_L \mathcal{T}$,

$$\begin{array}{ccc} M(X) & \xrightarrow{\lambda_{(M,H)}[h]} & N(X) \\ {\scriptstyle M(x)}\downarrow & \text{comm.} & \downarrow{\scriptstyle N(x)} \\ M(X') & \xrightarrow[\lambda'_{(M,N)}[h]]{} & N(X') \end{array} .$$

One verifies easily (in this spirit of starting a new life) that $\mathrm{Des}(I)$ is a Boolean pretopos in such a way that the forgetful functor

$$\begin{array}{lcl} P : \mathrm{Des}(I) & \longrightarrow & T \\ \quad (X, \lambda) & & X \\ \quad x \downarrow & \longmapsto & \downarrow x \\ \quad (X', \lambda') & & X' \end{array}$$

is a conservative pretopos functor (therefore, it "defines" the pretopos structure of $\mathrm{Des}(I)$). The functor J is given as

$$\begin{array}{lcl} S & \longrightarrow & \mathrm{Des}(I) \\ Y & & (IY, \boldsymbol{f}_Y) \\ y \downarrow & \longmapsto & \downarrow Iy \\ Y' & & (IY', \boldsymbol{f}_{Y'}) \end{array} ,$$

where $\boldsymbol{f}_Y \overset{=}{\text{def}} \boldsymbol{f}_{IY}\boldsymbol{Y} = \boldsymbol{Y}'$. In (2), R is defined as before, with $\boldsymbol{K} \overset{=}{\text{def}} \mathbf{Mod}^*(\mathcal{T}/L)$ the ultragroupoid defined in 10.(i). We have that $R = \boldsymbol{K}^*$, the dual of $\boldsymbol{K}$ in the sense of section 5 . The functor K is defined as evaluation of IFS's as actual functors:

$$\begin{array}{ccc} \mathrm{Des}(I) & \xrightarrow{K} & R \\ & & \\ (X, \lambda) & & [X, \lambda] : \mathrm{Mod}^*(\mathcal{T}/L) \to \mathrm{Set}^* \\ x \downarrow & \longmapsto & \downarrow [M \longmapsto M(x)] \\ (X', \lambda') & & [X', \lambda'] \end{array}$$

The fact that $K(x)$ as defined is indeed a natural transformation $K((X, \lambda)) \to K((X', \lambda'))$ is our definition of what an arrow $x: (X, \lambda) \to (X', \lambda')$ is. But the main point is that $K((X, \lambda))$ is indeed a strict ultra*functor, as we proved in sections 9 and 10 (see 10.(viii)).

The composite $KJ: S \longrightarrow R = \boldsymbol{K}^*$ is identical to the transpose of $F: \boldsymbol{K} \longrightarrow S^*$, where F is the full and faithful, relation-conservative ultra*functor of 10.(i); hence, by 8.2,

$F^{\#} = KJ$ is a quotient morphism. So, in particular, KJ covers R in the sense of 2.1.

It is obvious that K is a pretopos morphism reflecting isomorphisms. A key point is that K is full. For this, we need the following fact. With $(X, \lambda[\mathbf{f}])$ an IFS, and $h: M \xrightarrow{\cong} N$ an isomorphism of $\mathcal{T}$-models, we have

$$\lambda_{(M,N)}[h\restriction L]] = h_X \qquad (h\restriction L \overset{\text{def}}{=} \langle h_Y \rangle_{Y\in L}). \tag{4}$$

This is seen as follows. We have the isomorphism

$$(M, N, h|L) \xrightarrow{(h,\, 1_N)} (N, N, 1_{N|L})$$

as shown by the diagram

$$\begin{array}{ccc} M\restriction L & \xrightarrow{h\restriction L} & N\restriction L \\ {\scriptstyle h\restriction L}\downarrow & & \downarrow {\scriptstyle 1_N\restriction L = 1_{N\restriction L}} \\ N\restriction L & \xrightarrow[1_{N\restriction L}]{} & N\restriction L \end{array} .$$

But, the $\mathcal{T}+_L\mathcal{T}$-formula $\lambda[\mathbf{f}]$ is invariant under isomorphisms of $\mathcal{T}+_L\mathcal{T}$-models. This means that

$$\begin{array}{ccc} M(X) & \xrightarrow{\lambda_{(M,N)}[h\restriction L]} & N(X) \\ {\scriptstyle h_X}\downarrow & & \downarrow {\scriptstyle (1_N)_X = 1_{N(X)}} \\ N(X) & \xrightarrow[\lambda_{(M,N)}(1_{N\restriction L}) \;=\; 1_{N(X)}]{} & N(X) \end{array}$$

commutes, which shows what we want.

Now, the fullness of K goes as follows. We have IFS's (X, λ), (X', λ'), and ultra*transformation $\xi: [X, \lambda] \xrightarrow{\cong} [X', \lambda']$. This means, among others, that $\xi_M: MX \xrightarrow{\cong} MX'$, that is, $\xi_M: (\varepsilon_T(X))(M) \xrightarrow{\cong} (\varepsilon_T(X'))(M)$, with $\varepsilon_T: T \longrightarrow T^*$ the counit map for the adjunction. We claim that $\langle \xi_M \rangle_{M\in \mathrm{Mod}(T)}$ constitutes an ultratransformation $\varepsilon_T(X) \longrightarrow \varepsilon_T(X')$. The only problem is naturality. If $h: M \xrightarrow{\cong} N$ in $\mathrm{Mod}^*(T)$, we want

$$\begin{array}{ccc} MX & \xrightarrow{h_X = \lambda_{(M,N)}[h\upharpoonright L]} & NX \\ \xi_M \downarrow & & \downarrow \xi_N \\ MX' & \xrightarrow[h_{X'} = \lambda'_{(M,N)}[h\upharpoonright L]]{} & NX' \end{array}$$

to commute, with h_X, $h_{X'}$ in the horizontal positions. But by replacing the latter with their equals as shown (see (4)), we have an instance of the naturality of $\xi : [X, \lambda] \xrightarrow{\cong} [X', \lambda']$. This shows the claim; by invoking the fullness of ε_T (by 8.1), the fullness of K follows.

Given that K is a full and conservative pretopos functor, we immediately see that KJ covers R implies that J covers $\mathrm{Des}(I)$. The fact that J is full on subobjects is easily seen to be equivalent to BETH as stated in section 3. Thus, by 2.1, we have that J is a quotient morphism as we wanted.

What we have gotten rid of in this proof is the 2-categorical part, i.e., sections 2 and 11.

(3) is essentially the same as the descent theorem in the form of 3.1. Also, in fact, the co-regular factorization of I is

$$\begin{array}{ccc} S & \xrightarrow{\quad I \quad} & T, \\ & {\scriptstyle J}\searrow \quad \nearrow{\scriptstyle P} & \\ & \mathrm{Des}(I) & \end{array}$$

thus we have the theorem in the form of 2.2.

Incidentally, the argument for (4) is buried in section 2 where we should verify the 2-regular factorization for $\mathcal{CAT}$. The precise point is where, in

$$K \underset{P_1}{\overset{P_0}{\rightrightarrows}} A \xrightarrow{S} C, \quad C \xrightarrow{Z} D, \quad A \xrightarrow{H} D, \qquad \mu : HP_0 \longrightarrow HP_1,$$

with $Z = (q_D^*)^{-1}(H, \mu)$, we want to see that $Z \circ S = H$.

To close the paper, we describe (and answer) William Boshuck's observations concerning the papers [M1] and [M2]; one of these observations is that there is a mistake in [M2], the other that an argument in [M1], and hence in [M2] as well, can be simplified.

The mistake is in Definition 6.4 of cell-system in [M2]. To correct it, replace condition (0) there by the following:

(0) Λ_0 is the set of $\prec$-minimal elements of $|\Lambda|$; $\lambda_{\mathrm{in}} \in \Lambda_0$;

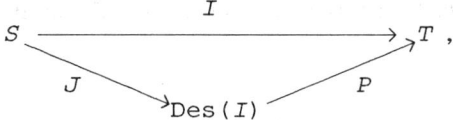

nothing else needs to be changed. The correction itself was also suggested by W. Boshuck.

In [M2], certain arguments are omitted, and they are said to be essentially the same as certain corresponding arguments in [M1]. One such is the proof of 6.10 in [M2]; this is said to be similar to a proof given in [M1]. It turns out that, with the definition of "cell-system" given in [M2], and with the change from a "pointed" context in [M1] to a (simpler) "unpointed" one in [M2], the latter proof cannot quite be transposed to [M2].

Note that in this paper, the "uncorrected" version of condition (0) in the definition of "cell-system" is used. However, we use, in contrast to [M2], "truncated" specifications (see 6.1).

The situation is understood when one considers definition 6.(iv), and the proof of 6.(v). With the non-truncated notion of ultraproduct specification, 6.(iv) needs a new clause, 6.(iv)(0):

G restricted to Λ_0 is a bijection onto $|\Gamma|$;

and 6.(iv)(2) should become:

If $v \in \mathcal{U}_\Gamma$, $P \in U_v$ and $g: P \longrightarrow |\Lambda|$ such that $G \circ g = g_v \upharpoonright P$, there is $u \in \mathcal{U}_{2\Lambda}$ such that $G(u) = v$ and $g_u \upharpoonright P = g$.

Note that the reference to an arbitrary $P \in U_v$ is essential in this clause; this can be seen when 6.(v) is applied in 7.(iii). When, in the proof of 6.(v) in Boshuck's version, an attempt is made to satisfy (2) by adding a suitable u, the value given to $g_u(i)$ for $i \in U_v - P$ will be "dummy" $\lambda \in \Lambda_0$ for which $G(\lambda) = G(g_v(i))$.

W. Boshuck's other observation (that he found when pursuing his own aims in a similar context) is the following simplified proof of 4.7 in [M1]. With Y, N and M as in the proof in [M1], and with Σ a subobject of Y restricted to $\boldsymbol{K}$ (we do not use here the "pointed" versions such as Σ^*, etc.), we want to show that $\Sigma(N)$ is determined from $\Sigma(M)$. Instead of using the Keisler-Shelah isomorphism theorem as in *loc.cit.*, we use a commutative diagram

$$\begin{array}{ccc} N & \xrightarrow{h} & M^U \\ {\scriptstyle\delta}\downarrow & \swarrow{\scriptstyle k} & \\ M^V & & \end{array}$$

with δ the canonical embedding, and h and k suitable elementary embeddings (the existence of such a diagram is elementary). Now, consider the following commutative diagram derived by applying Y and Σ :

$$\begin{array}{ccccc} YN & \xrightarrow{Yh} & YM^U & \xrightarrow{Yk} & YN^V \\ \uparrow & & \uparrow & & \uparrow \\ \Sigma N & \longrightarrow & \Sigma M^U & \longrightarrow & \Sigma N^V \end{array} ,$$

with vertical arrows inclusions; for simplicity, we have assumed that Y is a strict ultrafunctor.

Since Y preserves the canonical embedding δ, the outside quadrangle is the same as

$$\begin{array}{ccc} A & \xrightarrow{\delta_A} & A^V \\ \uparrow & & \uparrow \\ B & \xrightarrow[\delta_B]{} & B^V \end{array} ,$$

with $A = YN$, $B = \Sigma N$. Such a diagram (in Set) is always a pullback (as noted and used in [M2] in the corresponding proof of 7.3 there). Thus, in the previous diagram, the outside quadrangle is a pullback; therefore, since the middle vertical is a mono, so is the left-hand side square; this shows the desired assertion.

A similar simplification obtains in the context of [M2]; specifically, in the proof of 7.3.; in producing the starting commutative triangle, one uses 5.13 and 5.12 *loc.cit.*; 5.15 becomes superfluous (although it looks like a nice result ...).

In this case, this simplification becomes a real improvement; the statement of the "reconstruction result" 3.2 in *loc.cit.* becomes simpler inasmuch the references to (ω-type) colimits in the notions of ultragraph and ultradiagram can be deleted. These colimits were only used for the purposes of 7.3. Note that there was an anomaly in the generalization of the concept of ultra-category from [M1] to accessible pretoposes in [M2] because of the use of the ω-type colimits. The final goal in [M2] of an abstract codensity result was achieved, but at the cost of an unwelcome complication in the ultra-theory. In fact, the addition of colimits to the ultrastructure would destroy the possibility of the use M. Zawadowski, and later myself in this paper, have made of it. With the anomaly now removed from the "accessible" generalization, we have e.g. the direct generalization of 8.2' (mentioned in Section 8 of the present paper) to an accessible pretopos T and accessible models of it.

I do not see, however, the corresponding generalization for accessible *Boolean* pretoposes (generalizing 8.2). This is because Boshuck's simplification does not seem to apply in the Boolean case when all arrows allowed between models are *isomorphisms*; that is why, in this paper, at the end of section 8, we referred to the *original* proof of 7.3 of [M1]. It may be noted, on the other hand, that, because of the presence of ultralimits, the Keisler-Shelah isomorphism theorem can now be replaced by the (simpler) isomorphism theorem (due to S. Kochen) involving ultralimits.

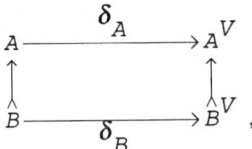

REFERENCES

[B] M. Barr, Exact categories. In: *Lecture Notes in Math.*, no. **226**, Springer-Verlag, 1972; pp. 1-120.

[Be1] J. Benabou, Introduction to bicategories. In: *Lecture Notes in Math.*, no. **47**, Springer-Verlag, 1967; pp. 1-77.

[B/J] A. Boileau and A. Joyal, La logique des topos. *J. Symbolic Logic* **46**(1981), 6-16.

[B/K/P] R. Blackwell, G. M. Kelly and J. Power, Two-dimensional monad theory. *J Pure and Applied Algebra* **59**(1989), 1-41.

[C/K] C. C. Chang and H. J. Keisler, *Model Theory*, third edition. North-Holland, 1990.

[CWM] S. Mac Lane, *Categories for the Working Mathematician.* Springer-Verlag, 1971.

[G] A. Grothendieck, Technique de descente et théorems d'existence en Géometrie Algébrique, I , II et III. *Séminaire Bourbaki,* N^{0} **190** (décembre 1959), N^{0} **195** (1959/60), et N^{0} **212** (février 1961).

[J] P. T. Johnstone, *Stone spaces.* Cambridge University Press, 1982.

[J/T] A. Joyal and M. Tierney, An extension of the Galois theory of Grothendieck. *Memoires Amer. Math. Soc.* **51**(1984), no. 309.

[K] H. J. Keisler, Limit ultraproducts. *J. Symbolic Logic* **30**(1965), 212-234.

[M/R] M. Makkai and G. E. Reyes, *First Order Categorical Logic.* Lecture Notes in Math., no. **611**, Springer-Verlag, 1977.

[M1] M. Makkai, Stone duality for first order logic. *Advances in Math.* **65**(1987), 97-170.

[M2] M. Makkai, Strong conceptual completeness for first order logic. *Annals of Pure and Applied Logic* **40**(1988), 167-215.

[M3] M. Makkai, Ultraproducts and categorical logic. In: *Methods in Mathematical Logic* (Proceedings, Caracas 1983, ed. C. Di Prisco), Lecture Notes in Math., no **1130**, Springer-Verlag, 1985; pp. 222-309.

[M4] M. Makkai, A survey of basic stability theory. *Israel J. Math.* **49**(1984), 181-238.

[M5] M. Makkai, A theorem on Barr-exact categories with an infinitary generalization. *Annals of Pure and Applied Logic* **47**(1990), 225-268.

[M6] M. Makkai, Full continuous embeddings of toposes. *Trans. Amer. Math. Soc.* **269**(1982), 167-196.

[M/P] M. Makkai and A. M. Pitts, Some results for locally finitely presentable categories. *Trans. Amer. Math. Soc.* **299**(1987), 473-496.

[M/Pa] M. Makkai and R. Paré, *Accessible Categories*. Contemporary Mathematics, vol. **104**, Amer. Math. Soc., 1989.

[P1] A. M. Pitts, Amalgamation and interpolation in the category of Heyting algebras. *J. Pure Appl. Algebra* **29**(1983), 155-165.

[P2] A. M. Pitts, An application of open maps to categorical logic. *J. Pure Appl. Algebra* **29**(1983), 313-326.

[P3] A. M. Pitts, Interpolation and conceptual completeness for pretoposes via category theory. In: *Mathematical Logic and Theoretical Computer Science* (Proceedings, Maryland 1985, ed. D. W. Kueker et al.), Marcel Dekker, 1987; pp. 301-328.

[P4] A. M. Pitts, Conceptual completeness for first-order intuitionistic logic: an application of categorical logic. *Annals of Pure and Applied Logic* **41**(1989), 31-81.

[S1] R. Street, Fibrations in bicategories. *Cahiers Top. Geom. Diff.* **21**(1980), 111-160.

[S2] R. Street, Two-dimensional sheaf theory. *J. Pure Appl. Algebra* **23** (1982), 251-230.

[S3] R. Street, Characterization of bicategories of stacks. In: *Lecture Notes in Math.*, no. **962**, Springer-Verlag, 1982; pp. 282-291.

[SGA4] M. Artin, A. Grothendieck and J. L. Verdier, *Theorie des Topos et Cohomologie Etale des Schemas*. Lecture Notes in Math., no's **269** and **270**, Springer-Verlag, 1972.

[Sh] S. Shelah, *Classification Theory*. North-Holland, 1978.

[TT] P. T. Johnstone, *Topos Theory*. Academic Press, 1977.

[Z1] M. Zawadowski, Un Theoreme de la Descente pour les Pretopos. *Ph. D. Thèse de doctorat*, Université de Montreal, 1989.

[Z2] M. Zawadowski, Descent and duality. To appear in *Annals of Pure and Applied Logic*.

Author's address: Department of Mathematics and Statistics, McGill University, 805 Sherbrooke Street West, Montreal, Quebec, Canada, H3A 2K6.

Editorial Information

To be published in the *Memoirs*, a paper must be correct, new, nontrivial, and significant. Further, it must be well written and of interest to a substantial number of mathematicians. Piecemeal results, such as an inconclusive step toward an unproved major theorem or a minor variation on a known result, are in general not acceptable for publication. *Transactions* Editors shall solicit and encourage publication of worthy papers. Papers appearing in *Memoirs* are generally longer than those appearing in *Transactions* with which it shares an editorial committee.

As of July 8, 1993, the backlog for this journal was approximately 8 volumes. This estimate is the result of dividing the number of manuscripts for this journal in the Providence office that have not yet gone to the printer on the above date by the average number of monographs per volume over the previous twelve months, reduced by the number of issues published in four months (the time necessary for preparing an issue for the printer). (There are 6 volumes per year, each containing at least 4 numbers.)

A Copyright Transfer Agreement is required before a paper will be published in this journal. By submitting a paper to this journal, authors certify that the manuscript has not been submitted to nor is it under consideration for publication by another journal, conference proceedings, or similar publication.

Information for Authors and Editors

Memoirs are printed by photo-offset from camera copy fully prepared by the author. This means that the finished book will look exactly like the copy submitted.

The paper must contain a *descriptive title* and an *abstract* that summarizes the article in language suitable for workers in the general field (algebra, analysis, etc.). The *descriptive title* should be short, but informative; useless or vague phrases such as "some remarks about" or "concerning" should be avoided. The *abstract* should be at least one complete sentence, and at most 300 words. Included with the footnotes to the paper, there should be the 1991 *Mathematics Subject Classification* representing the primary and secondary subjects of the article. This may be followed by a list of *key words and phrases* describing the subject matter of the article and taken from it. A list of the numbers may be found in the annual index of *Mathematical Reviews*, published with the December issue starting in 1990, as well as from the electronic service e-MATH [**telnet e-MATH.ams.org** (or **telnet 130.44.1.100**). Login and password are **e-math**]. For journal abbreviations used in bibliographies, see the list of serials in the latest *Mathematical Reviews* annual index. When the manuscript is submitted, authors should supply the editor with electronic addresses if available. These will be printed after the postal address at the end of each article.

Electronically prepared manuscripts. The AMS encourages submission of electronically prepared manuscripts in $\mathcal{A}_{\mathcal{M}}\mathcal{S}$-TEX or $\mathcal{A}_{\mathcal{M}}\mathcal{S}$-LATEX because properly prepared electronic manuscripts save the author proofreading time and move more quickly through the production process. To this end, the Society has prepared "preprint" style files, specifically the amsppt style of $\mathcal{A}_{\mathcal{M}}\mathcal{S}$-TEX and the amsart style of $\mathcal{A}_{\mathcal{M}}\mathcal{S}$-LATEX, which will simplify the work of authors and of the production staff. Those authors who make use of these style files from the beginning of the writing process will further reduce their own effort. Electronically submitted manuscripts prepared in plain TEX or LATEX do not mesh properly with the AMS production systems and cannot, therefore, realize the same kind of expedited processing. Users of plain TEX should have little difficulty learning AMS-TEX, and LATEX users will find that AMS-LATEX is the same as LATEX with additional commands to simplify the typesetting of mathematics.

Guidelines for Preparing Electronic Manuscripts provides additional assistance and is available for use with either AMS-TEX or AMS-LATEX. Authors with FTP access may obtain *Guidelines* from the Society's Internet node e-MATH@math.ams.org (130.44.1.100). For those without FTP access *Guidelines* can be obtained free of charge from the e-mail address guide-elec@math.ams.org (Internet) or from the Publications Department, American Mathematical Society, P.O. Box 6248, Providence, RI 02940-6248. When requesting *Guidelines*, please specify which version you want.

At the time of submission, authors should indicate if the paper has been prepared using AMS-TEX or AMS-LATEX. The *Manual for Authors of Mathematical Papers* should be consulted for symbols and style conventions. The *Manual* may be obtained free of charge from the e-mail address cust-serv@math.ams.org or from the Customer Services Department, American Mathematical Society, P.O. Box 6248, Providence, RI 02940-6248. The Providence office should be supplied with a manuscript that corresponds to the electronic file being submitted.

Electronic manuscripts should be sent to the Providence office immediately after the paper has been accepted for publication. They can be sent via e-mail to pub-submit@math.ams.org (Internet) or on diskettes to the Publications Department address listed above. When submitting electronic manuscripts please be sure to include a message indicating in which publication the paper has been accepted. No corrections will be accepted electronically. Authors must mark their changes on their proof copies and return them to the Providence office. Authors and editors are encouraged to make the necessary submissions of electronically prepared manuscripts and proof copies in a timely fashion.

Two copies of the paper should be sent directly to the appropriate Editor and the author should keep one copy. The *Guide for Authors of Memoirs* gives detailed information on preparing papers for *Memoirs* and may be obtained free of charge from AMS, Editorial Department, P. O. Box 6248, Providence, RI 02940-6248. For papers not prepared electronically, model paper may also be obtained free of charge from the Editorial Department.

Any inquiries concerning a paper that has been accepted for publication should be sent directly to the Editorial Department, American Mathematical Society, P. O. Box 6248, Providence, RI 02940-6248.

Editors

Recent Titles in This Series

(Continued from the front of this publication)

471 **Michael L. Mihalik and Steven T. Tschantz,** Semistability of amalgamated products and HNN-extensions, 1992
470 **Daniel K. Nakano,** Projective modules over Lie algebras of Cartan type, 1992
469 **Dennis A. Hejhal,** Eigenvalues of the Laplacian for Hecke triangle groups, 1992
468 **Roger Kraft,** Intersections of thick Cantor sets, 1992
467 **Randolph James Schilling,** Neumann systems for the algebraic AKNS problem, 1992
466 **Shari A. Prevost,** Vertex algebras and integral bases for the enveloping algebras of affine Lie algebras, 1992
465 **Steven Zelditch,** Selberg trace formulae and equidistribution theorems for closed geodesics and Laplace eigenfunctions: finite area surfaces, 1992
464 **John Fay,** Kernel functions, analytic torsion, and moduli spaces, 1992
463 **Bruce Reznick,** Sums of even powers of real linear forms, 1992
462 **Toshiyuki Kobayashi,** Singular unitary representations and discrete series for indefinite Stiefel manifolds $U(p,q;\mathbb{F})/U(p-m,q;\mathbb{F})$, 1992
461 **Andrew Kustin and Bernd Ulrich,** A family of complexes associated to an almost alternating map, with application to residual intersections, 1992
460 **Victor Reiner,** Quotients of coxeter complexes and P-partitions, 1992
459 **Jonathan Arazy and Yaakov Friedman,** Contractive projections in C_p, 1992
458 **Charles A. Akemann and Joel Anderson,** Lyapunov theorems for operator algebras, 1991
457 **Norihiko Minami,** Multiplicative homology operations and transfer, 1991
456 **Michał Misiurewicz and Zbigniew Nitecki,** Combinatorial patterns for maps of the interval, 1991
455 **Mark G. Davidson, Thomas J. Enright and Ronald J. Stanke,** Differential operators and highest weight representations, 1991
454 **Donald A. Dawson and Edwin A. Perkins,** Historical processes, 1991
453 **Alfred S. Cavaretta, Wolfgang Dahmen, and Charles A. Micchelli,** Stationary subdivision, 1991
452 **Brian S. Thomson,** Derivates of interval functions, 1991
451 **Rolf Schön,** Effective algebraic topology, 1991
450 **Ernst Dieterich,** Solution of a non-domestic tame classification problem from integral representation theory of finite groups ($\Lambda = RC_3, v(3) = 4$), 1991
449 **Michael Slack,** A classification theorem for homotopy commutative H-spaces with finitely generated mod 2 cohomology rings, 1991
448 **Norman Levenberg and Hiroshi Yamaguchi,** The metric induced by the Robin function, 1991
447 **Joseph Zaks,** No nine neighborly tetrahedra exist, 1991
446 **Gary R. Lawlor,** A sufficient criterion for a cone to be area-minimizing, 1991
445 **S. Argyros, M. Lambrou, and W. E. Longstaff,** Atomic Boolean subspace lattices and applications to the theory of bases, 1991
444 **Haruo Tsukada,** String path integral realization of vertex operator algebras, 1991
443 **D. J. Benson and F. R. Cohen,** Mapping class groups of low genus and their cohomology, 1991
442 **Rodolfo H. Torres,** Boundedness results for operators with singular kernels on distribution spaces, 1991
441 **Gary M. Seitz,** Maximal subgroups of exceptional algebraic groups, 1991
440 **Bjorn Jawerth and Mario Milman,** Extrapolation theory with applications, 1991
439 **Brian Parshall and Jian-pan Wang,** Quantum linear groups, 1991

(See the AMS catalog for earlier titles)